ILLIBERAL *Reformers*

Race, Eugenics & American Economics in the Progressive Era

ILLIBERAL *Reformers*

THOMAS C. LEONARD

PRINCETON UNIVERSITY PRESS
PRINCETON AND OXFORD

Published by Princeton University Press, 41 William Street,
Princeton, New Jersey 08540

In the United Kingdom: Princeton University Press, 6 Oxford Street,
Woodstock, Oxfordshire OX20 1TR

press.princeton.edu

Fourth printing, first paperback printing, 2017

Paperback ISBN: 978-0-691-17586-7

The Library of Congress has cataloged the cloth edition as follows:

Leonard, Thomas C., 1960– author.
Illiberal reformers : race, eugenics, and American economics in the Progressive era /
Thomas C. Leonard.
pages cm
Includes index.
ISBN 978-0-691-16959-0 (hardcover : alk. paper) 1. Economics—United States—History.
2. Progressivism (United States politics)—History. 3. Eugenics—United States—History.
4. United States—Economic conditions—1865–1918. 5. United States—Economic policy—To 1933. 6. United States—Social conditions—1865–1918. I. Title.
HC105.L46 2016
330.973'091—dc23
2015023243

British Library Cataloging-in-Publication Data is available

This book has been composed in Garamond Premier Pro

Printed on acid-free paper. ∞

Printed in the United States of America

5 7 9 10 8 6 4

CONTENTS

ACKNOWLEDGMENTS

A book, like any manufacture, is the product of many helping hands. An author stands in for the many others who have made intellectual and personal contributions. Every paragraph of *Illiberal Reformers* incurs a debt of some kind, and only a few can be acknowledged here.

Princeton University gave sabbatical time for research and writing. David Levy and Sandy Peart blazed the trail and are models of scholarly generosity. I am especially grateful for the advice and support of Malcolm Rutherford, who read the entire manuscript, as did Steve Medema. Both made the book better. Three anonymous referees for Princeton University Press gave the manuscript a scrupulously close and thoughtful reading. Their critical responses improved the book in significant ways, even if they would have told the story differently. Dan Rodgers long ago provided encouragement and read a portion of the final manuscript.

Parts of the book have been tried out in seminars over the years. Scholars at Princeton, Duke, New York University, George Mason, and many other places sharpened my arguments, both when they liked them and when they didn't. Peter Dougherty, my editor at Princeton University Press, gave the book its title and provided sage counsel. The wise and wonderful Carol Rigolot encouraged the project and is dearly missed at Princeton.

Some of the raw materials of *Illiberal Reformers* were published in the *History of Political Economy*. I thank *HOPE* for letting me revise, enlarge, and rework portions of: "More Merciful and Not Less Effective: Eugenics and Progressive-Era American Economics." *HOPE* 35(4): 709–734; "Mistaking Eugenics for Social Darwinism: Why Eugenics Is Missing from the History of American Economics." *HOPE* 37(supplement): 195–228; "American Economic Reform in the Progressive Era: Its Foundational Beliefs and Their Relation to Eugenics." *HOPE* 41(1): 109–141; and "Religion and Evolution in Progressive Era Political Economy: Adversaries or Allies?" *HOPE* 43(3): 429–469.

Illiberal Reformers I dedicate to four extraordinary Leonard women. My mother, Bonnie, inspired me to write it. My wife, Naomi, made the book possible. If her patience, wisdom, and generosity have bounds, I have not found them. This book was conceived when my daughters, Amara and Lilian, were girls. They are now young women, part of the present's dialogue with the past about the future, which is theirs.

PROLOGUE

Illiberal Reformers tells the story of the progressive scholars and activists who led the Progressive Era crusade to dismantle laissez-faire, remaking American economic life with a newly created instrument of reform, the administrative state. If many of their names are unfamiliar today, the progressives changed everything, permanently altering the course of America's economy and its public life.

American economic reform acquired its scientific and political authority during the turbulent economic times between the collapse of Reconstruction in 1877 and the US entry into the First World War in 1917. During these four decades, the period of the Gilded Age and the Progressive Era, the United States became a modern, urban, industrial, and multicultural world power, its spectacular rise propelled by an industrial revolution that transformed America. The vital national issues of the late nineteenth century—economic depression, financial panic, labor conflict, money wars, big business, immigration, and the tariff—were economic in nature, and public discourse placed economics at the center of a vigorous national debate over where and how government should respond to the consequences of an economic transformation that reached into the country's remotest corners.

Part I of *Illiberal Reformers* tells the story of the progressives' ascendancy, in three acts. In the first act, seizing the opportunity of recurring economic crisis, the progressives, many of them Protestant evangelicals on a self-appointed mission to redeem America, turn professional, finding different ways to make reform a vocation. The progressive economists completely remade the nature and practice of their own enterprise. From 1880 to 1900, both fostering and benefiting from a revolution in American higher education, the progressive economists established economics as a university discipline, transforming American political economy from a species of public discourse among gentlemen into an expert, scientific practice—economics.

Other economic progressives brokered their ideas in journalism, in reform organizations, in the community, and in public life. But they too were engaged with fundamentally economic questions—unemployment, low wages, long hours, workplace safety, industrial consolidation, immigration, and more—and they too undertook social investigations designed to produce economic knowledge and to influence public opinion and policymakers. The progressives gave us the professor of social science, the scholar-activist, the social worker, the muckraking journalist, and the economic expert advising or serving in government.

In the second act, the economic progressives forged the new authority of social science into rhetorical weapons, helping convince Americans and their political leaders that laissez-faire was both economically outmoded and ethically inadequate. Industrial capitalism, progressives said, created conflict and dislocation, operated wastefully, and distributed its copious fruits unjustly. Moreover, the new economy featured novel organizational forms—trusts, natural monopolies, industrial corporations, and industrial labor unions—and a rapidly increasing economic interdependence wrought by the furious pace of economic growth. Free markets, to the extent they ever could, no longer self-regulated. Progress, the economic progressives argued, now required the visible hand of a powerful administrative state, guided by expert social scientists—a model of economic governance progressives borrowed from scientific management.

In the third act, the economists joined their progressive allies in a crusade to reform and remake American government. If an administrative state were to be the new guarantor of economic progress, it would need to be built. By March 1917, the end of Woodrow Wilson's first term, it was. Countless additions would later be made to the new regulatory edifice, but the "fourth branch" of government was established.[1] The US government now directly taxed personal incomes, corporations, and estates. It dissolved prominent industrial combinations in steel, oil, tobacco, and sugar. Its new Federal Reserve regulated money, credit, and banking. Its new Federal Trade Commission supervised domestic industry, and its new Tariff Commission regulated international trade. State and federal labor legislation mandated workmen's compensation, banned child labor, compelled schooling of children, inspected factories, fixed minimum wages and maximum hours, paid pensions to single mothers with dependent children, and much more.

The establishment of the fourth branch marked an epoch-making change in the relationship of government to American economic life. It also shifted political authority within the state, moving power from the courts and political parties to the new independent agencies of the executive, and from judges and politicians to bureaucratic experts, who represented themselves as objective scientists above the political and commercial fray, administering progress for the good of all.

THE PROGRESSIVE PARADOX

The economic progressives fashioned the new sciences of society, founded the modern American university, invented the think tank, and blueprinted and framed the American administrative state. Progressives built these vital institutions of American life to carry out the twinned principles at Progressivism's core: first, modern government should be guided by science and not politics; and second, an industrialized economy should be supervised, investigated, and regulated by the visible hand of a modern administrative state. In so doing, they reconstructed American liberalism.

There was a price to be paid, however, a price *Illiberal Reformers* explores in its second half. Part II of *Illiberal Reformers* also has several acts, but each tells the same dark story—the campaign of labor reformers to exclude the disabled, immigrants, African Americans, and women from the American work force, all in the name of progress (Chapters 8, 9, and 10, respectively).

The progressives combined their extravagant faith in science and the state with an outsized confidence in their own expertise as a reliable, even necessary, guide to the public good. They were so sure of their own expertise as a necessary guide to the public good, so convinced of the righteousness of their crusade to redeem America, that they rarely considered the unintended consequences of ambitious but untried reforms. Even more so, they failed to confront the reality that the experts—no less than the partisans, bosses, and industrialists they aimed to unseat—could have interests and biases of their own.

Of course the experts *did* have interests and biases, which manifested most conspicuously in their responses to what was the supreme economic question of the day: is labor getting its due? Politically charged and analytically daunting, the "labor question" encompassed the most compelling economic issues of the Gilded Age and Progressive Era, and is taken up in Chapter 5.

Economic progressives either ignored the plight of African Americans during the brutal reestablishment of white supremacy in the Jim Crow South, or, as in the case of Woodrow Wilson, justified it. Progressive economists provided essential intellectual support to the cause of race-based immigration restriction, which, in the early 1920s, all but ended immigration from Asia and southern and eastern Europe. Such progressive exemplars as Richard T. Ely, John R. Commons, and Edward A. Ross promoted an influential theory known as *race suicide*, Ross's term for the notion that racially inferior immigrants, by undercutting American workers' wages, outbred and displaced their Anglo-Saxon betters.

The same theory—that so-called unemployable workers were innately disposed to accept lower wages—was readily adapted to apply to African Americans, the disabled, and women. The leading lights of American economic reform advocated regulation of workers' wages and hours to bar or remove the unemployable from employment, on the grounds that their inferior nationality, race, gender, or intelligence made their economic competition a threat to the American workingman and to Anglo-Saxon racial integrity.

It is important to understand that the progressive campaign to exclude the inferior from employment was not (merely) the product of an unreflective prejudice. Progressive arguments warning of inferiority were deeply informed by elaborate scientific discourses of heredity. Darwinism, eugenics, and race science recast spiritual or moral failure as biological inferiority and offered scientific legitimacy to established American hierarchies of race, gender, class, and intellect.

Economic progressives were profoundly influenced by Darwin and other evolutionists. Chapter 6 shows how the economic progressives (and their critics) drew deeply on evolutionary science's conceptions of heredity, progress, competition, selection, fitness, organism, and the role of human beings in controlling nature. Chapter 7 shows the uses economic progressives made of race science and eugenics, the social control of human breeding.

Among other things, biological ideas offered Progressivism a conceptual scheme capable of accommodating the great contradiction at the heart of Progressive Era reform—its view of the poor as victims deserving state uplift and as threats requiring state restraint.

This unstable amalgam of compassion and contempt helps explain why Progressive Era reform lent a helping hand to those it deemed worthy of citizenship and employment while simultaneously narrowing that privileged circle by excluding the many it judged unworthy. Progressive Era reform at once uplifted and restrained, and did both in the name of progress. In practice, only white men of Anglo-Saxon background escaped the charge of hereditary inferiority, and even members of this privileged group were condemned as inferiors when they, as with *The Jukes* and other "white trash" families studied by eugenicists, were judged deficient in intellect and morals.[2]

The roster of progressives who advocated exclusion of hereditary inferiors reads like a Who's Who of American economic reform. It includes the founders of American economics: Edward Bemis, John R. Commons, Richard T. Ely, Irving Fisher, Arthur Holcombe, Jeremiah Jenks, W. Jett Lauck, Richmond Mayo-Smith, Royal Meeker, Simon N. Patten, and Henry R. Seager.

They were joined by the founders of American sociology, Charles Horton Cooley, Charles Richmond Henderson, and Edward A. Ross; pioneering social-work professionals, such as Edward Devine, Robert Hunter, and Paul U. Kellogg; and leading Protestant social gospelers, such as Walter Rauschenbusch and Josiah Strong. University presidents, such as the University of Wisconsin's Charles Van Hise and Stanford's David Starr Jordan, vigorously advocated exclusion of hereditary inferiors, as did such political journalists as Hebert Croly, such jurists as Oliver Wendell Holmes, Jr., and many other progressive luminaries, not least US Presidents Theodore Roosevelt and Woodrow Wilson. Their causes varied, as did their justifications, but they all advocated the exclusion of immigrants, African Americans, women, and the disabled.

The progressives were not the only Progressive Era intellectuals to traffic in reprehensible ideas. Conservatives and socialists also drank deeply from the seemingly bottomless American wells of racism, sexism, and nativism, and they, too, borrowed evolutionary and eugenic ideas in support of their politics.

But the progressives command the historian's attention, because they prevailed. It was the progressives who fashioned the new sciences of society, founded the modern American university, invented the think tank, and created the American administrative state, institutions still at the center of American public life and still defined by the progressive values that formed and instructed them.

Eugenics and race science are today discredited. But the progressive vision of how to govern scientifically under industrial capitalism lives on. Expertise in the service of an administrative state, what progressives called social control, has survived the discredited notions once used to uphold it. Indeed, it has thrived.

PART I

The Progressive Ascendancy

Truth speaks to power in many different tones of voice.
—James A. Smith

1

Redeeming American Economic Life

ECONOMIC REVOLUTION

When American economic life transformed itself in the last quarter of the nineteenth century, the world had never seen anything like it. A furious expansion of railroad networks, fueled by government loans and land grants, opened a vast continental market. American business, powered by a transformative set of new production technologies, industrialized on a revolutionary scale. Interstate commerce grew so rapidly that hundreds of local clock conventions had to be replaced by a national system of standardized time in 1883.

In 1870, the last of the Civil War amendments to the US Constitution was ratified. Thirty-five years later, the US economy had *quadrupled* in size. American living standards had doubled. US economic output surpassed each of the German, French, and Japanese empires in the 1870s. It overtook the nineteenth century's global colossus, the British Empire, in 1916.

The industrial juggernaut propelled the American economy upward but did so undependably. Financial crises triggered prolonged economic depressions in the 1870s and the 1890s. Growth also distributed its copious fruits unevenly, creating vast industrial fortunes alongside disgruntled rural homesteaders and a newly visible class of the urban poor, a contrast journalist Henry George encapsulated as *Progress and Poverty*, a runaway best seller.

The transformation from an agricultural to an industrial economy—and from rural communities to a metropolitan society—produced social dislocations so unprecedented as to require new words, such as *urbanization*, a term coined in Chicago in 1888 to describe the migration from farm to factory and the explosive growth of America's industrial cities. Just over half of American workers in 1880 worked on farms. By 1920, only one-quarter remained on the land.[1] Crowded into tenements, urban workers confronted substandard housing, poor sanitation, and recurring unemployment.

Industry's voracious but volatile demand for labor was met by immigration to America on a grand scale, which introduced polyglot peoples with disparate cultural and religious traditions. Fifteen million immigrants arrived in the United States between 1890 and 1914, and nearly 70 percent of the new arrivals were Catholics, Jews, and Orthodox Christians from southern and eastern Europe. Most congregated in the cities. In 1900, three out of four people in New York City, Chicago, Boston, and San Francisco were immigrants and their children. By 1910, the foreign born accounted for 22 percent of the US labor force and for 41 percent of non-farm laborers.[2]

Industrialization and immigration gave rise to a labor movement whose growth was as fitful as the economy's. Labor unions grew explosively from 1880 to 1886, from a mere 168,000 to 1.2 million members. The violence of the 1886 riots in Chicago's Haymarket Square undid these gains. Organized labor then recovered its 1886 level in 1900, after which another surge doubled union membership to 2.4 million in 1904.[3]

Labor conflict was rampant and sometimes violent. From 1881 to 1905, American workers organized an average of four strikes per day, more than 36,000 in total.[4] Names like Homestead (1892), where steelworkers engaged in pitched battles with Carnegie Steel's armed strike breakers, and Pullman (1894), a strike that brought US railroads to a standstill until President Grover Cleveland deployed US Army troops to quash it, still commemorate the industrial violence of the era.

The turn of the century produced a new form of economic organization, the consolidated firm, or "trust." Between 1895 and 1904, a sweeping merger movement consolidated scores of American industries: 1,800 major industrial firms disappeared into 157 mergers. Nearly half of the consolidated giants enjoyed market shares of more than 70 percent.[5]

The new industrial behemoths were of a scale Americans could barely comprehend, 100 or even 1,000 times larger than the largest US manufacturing firms in 1870. John D. Rockefeller's Standard Oil Company was capitalized at $100 million in 1900. James Duke's American Tobacco Company reached $500 million in 1904, and the United States Steel Corporation was valued at $1.4 billion at its creation in 1901.[6]

Historian Thomas Haskell described the American economic transformation of the late nineteenth century as "the most profound and rapid alteration in the material conditions of life that human society has ever experienced."[7] Those who lived through it recognized its revolutionary aspects.

Simon Nelson Patten, a pioneering progressive economist at Pennsylvania's Wharton School, saw in industrialization an age of material abundance so unprecedented as to form a new basis for civilization. Wisconsin economist Richard T. Ely, the standard bearer of progressive economics, cofounded the American Economic Association in 1885 to organize and promote the new political economy required, he said, to comprehend a "new economic world." Frederick Jackson Turner told his fellow historians they were witnessing nothing less than the birth of a new nation. One can hardly believe, John Dewey marveled at the turn of century, "there has been a revolution in all history so rapid, so extensive, so complete."[8]

Patten, Ely, Turner, and Dewey were all progressive scholars making a case for economic reform, and none were strangers to hyperbole. But here they did not need to exaggerate. Conservative observers marveled no less at the speed and scope of the American industrial revolution. In 1890, David A. Wells, an influential Gilded Age defender of free trade and sound money, described the economic changes since the Civil War as the most important in all of human history.[9]

* * * * *

Revolution, which suggests abrupt discontinuity or rupture, is an imperfect term for changes wrought over forty years. But *revolution* is not inappropriate when we recognize that the late-nineteenth-century American economic transformation launched the United States on a permanently different economic course, with profoundly far-reaching and long-lived consequences. Between the end of Reconstruction and the United States' entry into the First World War, the speed and scope of economic change was such that few Americans could be spectators only. Welcome or not, change was thrust on them, and there was no choice but to meet it.

Ordinary Americans met economic change with responses as different as their situations. Some responded by embracing new opportunities, freedoms, and identities. Middle-class women went to work outside the home, glimpsing the prospect of greater economic independence and, for some, even a vocation other than motherhood. Young people found the new pleasures of city life liberating. Former journeymen started their own businesses, and some met with success. University enrollments more than quadrupled, giving women and a burgeoning middle class their first chance at higher education. Immigrants did not find streets paved with gold, but many found refuge from starvation, pogroms, and peonage.

For other Americans, change offered not new opportunities but new constraints, not new freedom but new oppression, not new identities but new stigmas. The brutal reestablishment of white supremacy in the American South confronted African Americans with disenfranchisement, debt peonage, and organized racial terror. Native Americans, decimated when Europeans colonized America, were decimated again by coerced relocations, carried out by a postbellum US Army in need of new missions. Egged on by agitators like Dennis Kearney, white mobs attacked Chinese immigrants, accusing them of undercutting the American workingman.

Hard money and deflation punished farmers and other debtors. When they joined the migration to the cities, farmers and journeymen discovered their hard-won skills mattered less. They might command higher compensation at the factory, but employment threatened their republican self-identities. Having been raised to disdain the "hireling," they now accepted wages themselves. A boss told them what to do, and did not care whether his factory hands had once owned land or other property.

Those disenfranchised, damaged, and devalued during the Gilded Age met change individually and also collectively. Farmers formed cooperatives, skilled workers organized trade unions, men joined fraternal groups, women started clubs, and immigrant communities created a host of mutual aid societies, which provided credit, insurance, and other mutual services. Evangelicals founded youth associations, the Salvation Army, and other agencies organized to redeem the impressionable and the fallen.

Activists such as Ida Wells exposed mob violence against African Americans and organized antilynching campaigns, at home and abroad. African Americans chose to leave the South's racial caste system, their migration northward quickened by job opportunities created during mobilization for the First World War.

These grassroots movements were an essential part of America's many and varied responses to the economic, social, and political consequences of industrialization. American historical writing began telling the stories of ordinary Americans in the 1970s. Before this historiographic turn, Progressive Era histories were political and focused on those who made reform a vocation—the progressives. It is their story that *Illiberal Reformers* tells.

THE ECONOMIC PROGRESSIVES

The longstanding emphasis on politics and reform professionals was itself a progressive legacy. The earliest accounts of Progressivism, written by such historians as Benjamin Parke DeWitt, were self-portraits.[10] They painted ordinary people into the background as passive victims of the rough winds of economic change. The progressives filled the foreground, a vanguard of selfless scholars and activists leading the People—if not any recognizable people—in a crusade against wealth and privilege.

To conceptualize the period as Progressive was to define it by its politics and to associate Progressivism with an elite class: political figures like Theodore Roosevelt and Woodrow Wilson, university social scientists, settlement-house workers, muckraking journalists, conservationists, Prohibitionists, and birth controllers.[11] The protest of the progressives originated not out of personal suffering but rather out of moral and intellectual discontent with the suffering (and enrichment) of others.[12]

Progressives did not work in factories; they inspected them. Progressives did not drink in saloons; they tried to shutter them. The bold women who chose to live among the immigrant poor in city slums called themselves "settlers," not neighbors. Even when progressives idealized workers, they tended to patronize them, romanticizing a brotherhood they would never consider joining.[13]

The distance progressives placed between themselves and ordinary people was not the product of class prejudice alone. Some progressives came from privilege, but far more were children of middle class ministers and missionaries, a number of whom struggled before finding vocational outlets for their intellectual and reform energies. The few who had known real deprivation, such as Thorstein Veblen, never romanticized it.

The distance progressives placed between themselves and ordinary people instead had its origins in the progressives' self-conception as disinterested agents of reform. As they devised ways to make reform a vocation, the progressives found themselves poised between the victims and the beneficiaries of economic transformation. Most opted not to choose sides. Instead, they portrayed themselves as the representatives of the common good, uniquely positioned to transcend personal, class, regional, and partisan interests.

If progressives agreed that they represented the common good, they regularly disagreed on what the common good was. W.E.B. Du Bois and Woodrow Wilson, for example, held entirely opposed views of the proper role of whites and blacks in American life.[14] Senator Robert La Follette vigorously opposed American entry into the First World War, while his one-time Wisconsin compatriot, progressive economist Ely, accused him of aiding the enemy.[15]

Ely and his University of Wisconsin colleagues, John R. Commons and Edward A. Ross, campaigned to bar immigrants they judged racially inferior, while other progressives, such as settlement-house worker Grace Abbott, upheld the America tradition of openness to newcomers, as we shall see in Chapter 9. The same trio of Wisconsin academics crusaded against the evils of alcohol, while John Dewey believed progressives had causes more important than the saloon. Theodore Roosevelt preferred to regulate the trusts, while "the people's lawyer," Louis Brandeis, wanted to break them up, as discussed in Chapter 4.

The upshot was a pattern of conflict and cooperation that led to shifting political alliances and to a reputation for fractiousness. "The friends of progress," Benjamin Parke DeWitt lamented in 1915, "are frequently the enemies of each other."[16]

As diverse and fractious as Progressive Era reformers could be, they all drew on a shared, recognizable, and historically specific set of intellectual understandings, what Daniel Rodgers has termed "discourses of discontent."[17] First, progressives were discontented with liberal individualism, which evangelicals called un-Christian, and more secular critics scorned as "licensed selfishness."[18] As we shall see in Chapter 2, the progressives were nationalist to the core, though they reified the collective using many names besides *nation*, such as the state, the race, the commonweal, the public good, the public welfare, the people, and, as discussed in Chapter 6, the social organism.[19] Whichever term they used, progressives asserted the primacy of the collective over individual men and women, and they justified greater social control over individual action in its name.

Second, progressives shared a discontent with the waste, disorder, conflict, and injustice they ascribed to industrial capitalism. The furious pace of change had produced unprecedented economic volatility and social dislocation. Many believed the remedy was improved efficiency, the quintessentially progressive idea that the application of science, personified by the efficiency expert or social engineer, could improve virtually any aspect of American life, Efficiency, in business and public administration, is the story of Chapter 4.

Monopoly describes the third source of progressive discontent. Industrial capitalism had brought forth unprecedented and gigantic forms of economic organization—trusts, pools, and combinations. Antimonopoly rhetoric comprised a host of objections to big business—destruction of small business, monopoly profiteering, unfair trade practices, deskilling of labor, exploitation of workers—joined with the longstanding republican fear that centralized economic power corrupted politics.[20]

Progressives used the language of anti-individualism, efficiency, and antimonopoly for varying purposes. But nearly all progressives used this rhetoric. And nearly all agreed, moreover, that the revolutionary consequences of industrial capitalism required rethinking and reforming American economic life and its governance. As Ely put it, laissez-faire was not only morally unsound, it was economically obsolete, a relic of a bygone era.[21] Whatever free markets had once accomplished, they now produced inefficiency, instability, inequality, and a tendency toward monopoly.

Few progressives were content merely to deplore the diseases of a modern industrial economy. America needed, they agreed, a new form of government, one that was disinterested, nonpartisan, scientific, and endowed with discretionary powers to investigate and regulate the world's largest economy, as well as to compensate those exploited, injured, or left behind—the administrative state.

Nothing was more integral to Progressivism than its extravagant faith in administration. The visible hand of administrative government, guided by disinterested experts who were university trained and credentialed, would diagnose, treat and even cure low wages, long hours, unemployment, labor conflict, industrial accidents, financial crises, unfair trade practices, deflation, and the other ailments of industrial capitalism. Chapter 3 tells the story of how a small band of scholars remade the nature and practice of their discipline, transforming themselves into expert economists in the service of the administrative state.

The progressives had different and sometimes conflicting agendas. But nearly all ultimately agreed that the best means to their several ends was the administrative state. In this crucial sense, Progressivism was less a coherent agenda of substantive goals that it was a technocratic theory and practice of how to obtain them in the age of industrial capitalism. The heart of Progressivism, as historian Robert Wiebe famously summarized it, was its ambition to "fulfill its destiny through bureaucratic means."[22]

* * * * *

Illiberal Reformers tells the story of the progressive scholars and activists who enlisted in the Progressive Era crusade to dismantle laissez-faire and remake American economic life through the agency of an administrative state. Historians, just like everybody else, work with the tools they have at hand. I am a historian of economics, and *Illiberal Reformers* shines its narrative lamp on the progressive economists. But this is not their story alone, and had it been, they would not have recognized it.

American economic reform in the Gilded Age and Progressive Era featured a large, eclectic, and sometimes fractious cast. Most would not have called themselves economists, but nearly all were engaged with fundamentally economic questions—unemployment, low wages, long hours, workplace safety, industrial consolidation, immigration, and more. All of them, not just the academics, undertook social investigations designed to produce economic knowledge and to influence public opinion and policymakers.

They inspected factories; mapped city slums; compiled wages and working hours for legal briefs; exposed corruption in government and malfeasance in business; did casework for scientific charity organizations; practiced scientific management, calculated family budgets and tax revenues; and measured the bodies and intellects of immigrants, schoolchildren, and Army recruits. *Everything*, as Jane Addams said, could be improved.

The progressives in economic reform were intellectuals with graduate schooling, and many had training in political economy. But most chose to pursue their reform vocations outside the universities, brokering their ideas in reform organizations, in journalism, in the community, and in public life.

The new research universities, exemplified by Johns Hopkins University (1876), were founded not to reproduce their faculties but to send civic-minded men and women into the world so they might improve it. The path-breaking graduate seminar in Historical and Political Science at Hopkins, directed by historian Herbert Baxter Adams and political economist Richard T. Ely, produced many talented scholars, but the University was no less pleased with the careers of Woodrow Wilson or journalist Albert Shaw. Graduates who remained in academia, such as Edward A. Ross and John R. Commons, were the antithesis of the cloistered scholar. They were public figures who threw themselves into economic and social reform, as they were expected to do.

The progressive economists made alliances; formed associations; and shared ideas, offices, and personnel with many other scholars and activists. Their progressive allies and colleagues included figures such as sociologists Charles Cooley, Albion Small, and Charles Richmond Henderson; ministers of the social gospel Washington Gladden and Lyman Abbott; settlement-house workers Jane Addams and Florence Kelley; labor reformer Josephine Goldmark; efficiency expert Frederick Winslow Taylor; municipal reformers Edward Bemis and Frederick Cleveland; scientific charity leader Edward T. Devine; social surveyor Paul U. Kellogg; journalists Albert Shaw and Walter Weyl; lawyers Louis Brandeis and Felix Frankfurter; and reform-minded politicians Robert La Follette, Theodore Roosevelt, and Woodrow Wilson, among others.

Some economic reformers were in reform organizations, some in the university, some in the community, some in public life, and some in all four. All were intellectuals that had turned off the expected scholarly path of the classics, theology, and philosophy to study the new social disciplines created to put reform into action—economics, politics, sociology, and public administration. They followed different paths to different places, but all of the progressives found a way to make a vocation of reform.

REDEEMING AMERICAN ECONOMIC LIFE

The first generation of progressive scholars and activists was born largely between the mid-1850s and 1870. Unlike the generation of 1840, which included such members as Henry George; Oliver Wendell Holmes, Jr.; William Graham Sumner; and Lester Frank Ward, the progressives were too young to have served in the Civil War.[23]

Nearly all descended from old New England families of seventeenth-century Massachusetts Bay background, families that, like America itself, had gradually moved westward. More often than not, progressives were the children of Protestant ministers or missionaries, fired with an evangelical urge to redeem America. The sons were expected to continue the family calling, and the daughters were expected to stay home, and both wanted neither.

The progressives' urge to reform America sprang from an evangelical compulsion to set the world to rights, and they unabashedly described their purposes as a Christian mission to build a Kingdom of Heaven on earth.[24] In the language of the day, they preached a social gospel.

The term *social gospel* describes a late-nineteenth-century and early-twentieth-century form of liberal Protestantism that pursued economic and social improvement through a scientifically informed mission of social redemption. It originated in liberal Protestantism's efforts to reckon with radically changed socioeconomic conditions and with modern scientific investigations into the origins of humankind and of Christianity's sacred texts.[25]

At the collapse of Reconstruction, American Protestant churches were no force for economic reform. The same was largely true of American political economy. The best-selling text in the second half of the nineteenth century was Arthur Latham Perry's *The Elements of Political Economy*, which taught students that providential design explained the remarkable capacity of free markets to promote the good of all.[26] Social mobility made America doubly blessed. "There is nothing to hinder any laborer from becoming a capitalist," Perry wrote, "nearly all our capitalists were formerly laborers."[27]

When T. E. Cliffe Leslie surveyed American political economy for his English readers in 1880, he described it as sectarian, and he scorned Perry's treatise as little more than a Sunday School catechism. But American Protestantism, like American political economy, utterly transformed its relationship to economic reform.

The American Economic Association (AEA), founded in 1885, embodied the social gospel's distinctive amalgam of liberal Protestant ethics, veneration of science, and the evangelizing activism of pious, middle-class reformers.[28] Clergyman Josiah Strong (1847–1916), author of the best-selling *Our Country*, an exaltation of Protestant Anglo-Saxon manifest destiny, praised the AEA for its Christian political economy.[29] Of the AEA's fifty-five charter members, twenty-three were clergymen, many of them national leaders of the social gospel movement, including Washington Gladden and Lyman Abbott.[30]

Richard T. Ely, the prime mover behind the AEA's establishment, exemplified the social gospel view of economic reform. The good Christian should be concerned with this world, Ely said, not with the next. The good Christian must go among the poor, as had Christ, lifting up even the most degraded by providing them personal contact with "superior natures."[31] The economic reformer's calling was to "redeem all our social relations," Ely declared, by establishing an earthly kingdom of righteousness.[32]

The AEA economists were young. Ely was thirty-one years old at time of the AEA's founding. Woodrow Wilson, a recent graduate student of Ely's, was twenty-eight, and had just begun his academic career at Bryn Mawr College. Edward W. Bemis, another newly minted Ely student, was twenty-five, then at Amherst College. Edwin R. A. Seligman, freshly appointed at Columbia College, was twenty-four. Among the senior charter members, John Bates Clark was, at thirty-eight, the oldest by four years.

In redirecting American Protestantism from saving souls to saving society, the social gospelers enlarged and transformed the idea of Christian redemption. John R. Commons, an Ely protégé who rose to the front ranks of progressive economics, affirmed the social gospel view that society was the proper object of redemption.[33]

Just as salvation was increasingly socialized, so too was sin. Edward A. Ross, like Commons, was a student of Ely's at Johns Hopkins who became a leading public intellectual of American Progressivism. Ross's *Sin and Society* summarized the view that sin was no longer a matter of inborn immorality. Sin, Ross wrote, was social in cause.[34]

Redeeming America required more than a reformed church. Social gospelers built an impressive network of voluntary agencies to encourage Christian betterment: Christian youth associations, Christian summer camps, the Salvation Army, immigrant settlement houses, and a host of other organizations intended to redeem to the impressionable, the fallen, and the newly arrived.[35] Ultimately, however, the social gospel economists, like all progressives, turned to the state.

Arthur Latham Perry had seen the hand of God in the way free market exchange benefited all. The social gospel economists, who opposed free markets but not divine purpose, relocated Him to the state. "God works through the State," Ely professed, more so than through any other institution, including, he implied, the church.[36] Commons told his Christian audiences that the state was the greatest power for good that existed among men and women.[37]

The AEA's intellectual leaders—Henry Carter Adams, John Bates Clark, and Simon Nelson Patten—were not quite as outspoken as Ely and Commons, but they too understood economic reform as a method of redeeming American economic and political life. Adams was born in 1851 in the frontier state of Iowa. His father, Ephraiam Adams, was a Congregationalist missionary who had moved his family to the wilderness so that they could dedicate their lives to building a Christian commonwealth west of the Mississippi.[38]

Henry intended to follow in his father's footsteps, enrolling at Andover Theological Seminary in 1875. But when he could not be born again, Henry abandoned the ministry for political economy. Economics, Adams wrote to his father upon leaving the seminary, was "work of a lower order than dealing directly—profoundly—with the souls of men, but it is work which a follower of Christ may do."[39] As Edwin R. A. Seligman remarked when memorializing him, economic reform was just a different path to Adams's original end, the redemption of America.[40]

Most of the AEA's intellectual leaders made a similar journey. John Bates Clark planned to enter the ministry until his Amherst College mentor, Julius Seelye, persuaded him to study political economy instead of enrolling at Yale Divinity School.[41] John R. Commons's mother, Clara Rogers, expected John to become a minister. He did not, finding his reform calling in economics. Recalling his graduate student days at Johns Hopkins, where Ely instructed him to do case work for the Baltimore Charity Organization Society, Commons said that being a social worker as well as a graduate student in economics was his "tribute to her longing that I should become a minister of the Gospel."[42]

Edwin R. A. Seligman, scion of a prominent German-Jewish banking family in New York, was the only Jew among the AEA charter members. But Seligman also sought refuge from the constraints of his religious inheritance, becoming an active supporter of his colleague Felix Adler's Society of Ethical Culture. No less than his social gospel colleagues, Seligman was impelled by a felt ethical obligation to improve the conditions of American economic life.

The social gospel claimed adherents in all the fledging American social sciences. The founder of the United States' first sociology department, Albion Small, was a graduate of the Newton Theological Seminary and a social gospeler.[43] His sociology colleague at the University of Chicago, Charles Richmond Henderson, was a minister who served as the university's chaplain. Just as Ely regarded the state, so Henderson regarded the new social sciences, as a God-given instrument of Christian economic reform. To aid the reformer, Henderson wrote in 1899, "God has providentially wrought out for us the social sciences and placed them at our disposal."[44]

The social gospel also deeply informed the pioneering social work of progressives like Jane Addams of Chicago's Hull House, who "settled" in poor urban neighborhoods to live among the dispossessed. Christianity, Addams said, was not a set of doctrines, but something immanent in humanitarian efforts to uplift fellow human beings, to find good in even the meanest places.[45]

The social gospel reformers, as postmillenarians, believed that a Kingdom of Heaven on earth could be built without Christ's return. Christian men and women, providentially equipped with science and the state, would build it with their own hands. In other words, the social gospelers believed they already held the blueprints for social and economic redemption.

The task of the social gospel reformer was that of the preacher—not merely to serve the social good but also to identify it for others. In Richard T. Ely's formulation, the economic reformer consciously adopts an ethical ideal, shows how it was be attained, and "encourage[s] people to strive for it."[46] Redemption required more than providing the poor with what they wanted but lacked; it required teaching the poor what they *should* want.

The social gospel went into decline during the First World War. The Great War's slaughter and uncontrolled irrationality mocked the progressive idea of spiritual and social progress through enlightened social control. But social gospel economics also suffered from developments internal to American social science. By the outbreak of the war in Europe in 1914, American economics had become an expert, scientific discipline, establishing a beachhead in the universities by 1900 and in government soon thereafter. Between 1900 and 1914, the imperatives of professionalization pushed progressives toward an economics less encumbered by social gospel pieties.[47]

Professional economics' turn away from the crusading language and imagery of the social gospel was neither sudden nor solely a matter of maintaining its scientific and professional bona fides.[48] The social gospelers recognized that the growing diversity of American Progressivism made their vision of a Protestant Christian commonwealth too sectarian. Catholics, Jews, Orthodox Christians, and others, millions more of whom had arrived on US shores between 1890 and 1914, held rather different views of what religious ethics demanded of the state.

When they recast their evangelical language in a more secular form, the economic progressives fashioned a discourse of an ethical science in the service of society. But even as they secularized their Christian idiom, they did not abandon the evangelical idealism driving their reform mission. Instead, they reconstituted it, making the social gospel into what historian David Hollinger has called the "intellectual gospel." The intellectual gospel represented scientific inquiry as itself a kind of religious calling, found religious potential in science, celebrated science in a religious idiom, and believed that "conduct in accord with the ethic of science could be religiously fulfilling."[49]

The progressives venerated science not only because it was their necessary instrument of social improvement. For the social gospel progressives at the forefront of American economic reform, science was a place of moral authority where the public-spirited could find religious meaning in scientific inquiry's values of dispassionate analysis, self-sacrifice, pursuit of truth, and service to a cause greater than oneself.

2

Turning Illiberal

The elder statesman the young progressive economists tapped for the first AEA presidency was Francis Amasa Walker, then forty-five. A member of the pre-progressive Generation of 1840, Walker missed a key formative experience of the AEA's founding core: graduate education in Germany. To obtain advanced instruction in political economy, which was all but unavailable from the sleepy, postbellum American colleges, American reformers traveled to the German universities, which, in the 1870s and early 1880s, were regarded as the world's finest in political economy.[1]

Germany exposed the young Americans to the ideas of the German Historical School of political economy, with its positive view of state economic intervention, quintessentially compulsory insurance against sickness, industrial accidents, debility, and old age. Most of their German professors—most influentially, Adolph Wagner and Gustav Schmoller of the University of Berlin, Johannes Conrad of the University of Halle, and Karl Knies of Heidelberg—were hostile to the idea of natural economic laws, which they disparaged as "English" economics, a swipe at the classically liberal tendency of political economy in Great Britain.[2]

Richard T. Ely noted that the young Americans quickly abandoned "the dry bones of orthodox English political economy for the live methods of the German school."[3] To understand contemporary economic life, the Germans said, you needed to know each nation's past, a study they called *Nationalökonomie*. If an economy was the path-dependent product of a nation's unique development, then its workings were not unalterable natural laws, they were historically contingent and subject to change. The right kind of government could control and shape a nation's economic life.

The German economists, though sometimes mocked as "socialists of the lectern," were no friends of revolutionary socialism. They successfully positioned their welfare statism as a reformist *via media* between the extremes of English laissez-faire and proletarian dictatorship. Ely made clear the German academics were not socialists in the "vulgar" sense of the term, referring to the revolutionaries of Germany's Social Democratic Party. The "socialists of the lectern" were, rather, socialists in a broader sense, men who believed social problems could no longer be left to individuals or voluntary associations, but must be dealt with by an expert-guided state.[4]

Perhaps even more influential than the German professors' statism and historicist method was their professional status. Wilhelmine Germany accorded its university professors political and scientific authority, enough that they formed part of the German "mandarin" class, an intellectual elite whose authority derived from their special education and their influence in the upper reaches of the civil service. In the mandarins' self-conception, they were not mere bureaucrats but an elite uniquely capable of transcending politics and objectively identifying the public good.[5] The Kaiser himself, Wilhelm I, attended Schmoller's lectures.[6] Harvard's Frank Taussig (1859–1940), among the most distinguished of America's Progressive Era economists and a man not prone to gushing, recalled his Berlin professors as having achieved "a degree of perfection ... that astonishes the world."[7]

German university professors in the 1870s and early 1880s occupied a social space as yet nonexistent in the United States: the academic scientist of society, with political influence and social standing. The American graduate students who traveled to Germany met there a new and compelling idea: economic reform could be a vocation, even a distinguished one.

PROFESSIONALIZING BY PROFESSING ECONOMICS

The progressives returned from Germany with their evangelical zeal to redeem America still hot if now tempered by the latest ideas in political economy and informed by a working model of economic reform. But their German professors' enviable scientific and political authority had no American equivalent whatsoever.

In 1880, the United States had *three* faculty members at the leading schools devoting most of their time to political economy. The prospects for making a vocation of economic reform in the university looked dim. But they brightened, and rapidly.[8]

The same industrial revolution that roused the Christian consciences of the progressives poured a geyser of new money into American universities, producing another revolution, this one in higher education. Capitalists' industrial fortunes lavishly endowed new universities built from scratch: Ezra Cornell (1868), Cornelius Vanderbilt (1871), Johns Hopkins (1876), Jonas Clark (1889), Leland Stanford (1891), and John D. Rockefeller (Chicago 1891).[9]

The doubling of American living standards made college accessible to a larger pool. Between 1870 and 1900 the number of students at American universities quadrupled, as did the ranks of American faculty members. Graduate student enrollments increased explosively, from 50 to nearly 6,000.[10] In 1903, William James could refer to the "PhD Octopus."[11]

The new universities threw off the cleric-dominated classics curriculum of traditional American college instruction and adopted the investigatory spirit and methods of the modern German universities. They offered more practical courses, notably political economy. Hopkins and Chicago broke all precedent by giving priority to graduate research and teaching over undergraduate instruction. The new research universities' intellectual example, and also their competition, spurred the colonial era colleges to catch up.

The growth of instruction in political economy can be measured against the classics, the longstanding foundation of the traditional college curriculum with which the upstart social science disciplines were vying. In 1880, college courses in Latin outnumbered courses in political economy by ten to one. By 1890, however, the ratio had decreased to three to one, and by 1900, it was down to two to one. At leading schools in 1900, there was parity. By 1912, only English had more undergraduate majors than did economics at Yale University.[12]

The course of American political economy's establishment as an academic discipline was tracked by the increasing currency of its new name, "economics." The name was exceedingly rare in university catalogs and other literature in the 1870s. "Political economy" predominates well into 1890s. But by 1900, "economics" had displaced the older term altogether.[13]

At the time of Civil War, professional social science simply did not exist. There were, of course, scholars engaged in social inquiry. But as Thomas Haskell argued, their inquiry was not professional, in the sense that it was neither specialized nor "decisively oriented to any ongoing, disciplined community of inquiry."[14]

The economists built their disciplinary community in the academy in relatively short order, from roughly 1880 to 1900. They led graduate students in German-style research seminars, granted PhDs, and wrote as specialists for one another in a half-dozen scholarly journals they founded and edited.[15] Textbooks setting out the new domain of American economics, such as Richard T. Ely's *Introduction to Political Economy* (1889), quickly appeared.[16]

The 1880–1900 revolution in American higher education established economics as a university discipline, transforming American political economy from a species of public discourse among gentlemen into an expert, scientific practice—economics. As A. W. Coats judged it, the late-nineteenth-century revolution in American higher education benefited no group as much as it did the economists.[17]

The American university gave the economists more than academic chairs, a decent library, and students. The American university gave the economists scientific authority—a gift not elsewhere obtainable and one that was essential to the progressives' mission of scientifically redeeming American economic life.

The economists understood that economics, like all the infant social sciences, needed the protection of the university. The economists who became academic leaders used their institutional clout to secure a permanent place for economics in the American university. Seven of the AEA's fifty-five charter members presided over major universities, including Brown, Cornell, Illinois, the Massachusetts Institute of Technology (MIT), Northwestern, Princeton, Wisconsin, and Yale.[18]

The economists professionalized, as all professions do, in the hope of monopolizing authority over a given area of knowledge and practice. The AEA was organized not merely to arrange scholarly meetings and promote the field. The AEA was formed to exclude other claimants to economic knowledge by making them outsiders and amateurs.

But unlike medicine, law, and engineering, the economists failed to enlist government to enforce the disciplinary boundaries they wanted recognized. It was illegal to practice medicine, law, or engineering without state permission, but anyone could lawfully practice economics without training or a license. So even though the economists successfully organized their learned society, they still needed the university to recognize and legitimize their claims to scientific authority.[19]

The change in the status of economists—from gentlemen amateurs to specialized professionals, from public intellectuals to scientific advisors—was rapid. All but nonexistent as a field of academic inquiry before 1870 and merely nascent by 1880, American economics was, by the turn of the century, established in the new university order.

Looking back from 1925, Princeton's Frank Fetter, speaking at an occasion honoring Richard T. Ely, marveled at the economists' accomplishments. It would be hard to find anywhere in the history of scholarship, Fetter claimed, "a higher average of success and achievement than this little band of pioneers attained."[20] Discounting for ceremonial hyperbole, Fetter had a point. The first generation of progressive economists had created, essentially *ex nihilo*, two new and influential vocations in America: the professor of economics and the expert economist in the service of the administrative state.

THE END OF THE AUTONOMOUS INDIVIDUAL

The progressive economists, like all educated Americans of their generation, had been weaned on Anglo-American individualism, with its natural-rights foundation. In the classically liberal model, a well-ordered society channeled self-interested market behavior into socially beneficial outcomes. Economic progress was not planned; it was a natural by-product of a healthy commercial society. Government's function was limited to ensuring an institutionally healthy environment for mutually beneficial trade.

The progressives called it laissez-faire. As Christians they judged laissez-faire to be morally unsound, and as economists they declared it functionally obsolete, a quaint relic now buried by the realities of Gilded Age capitalism. A political economy written before the introduction of railroads, Ely wrote, "can scarcely be sufficient in the year 1885."[21]

The progressives' German professors had taught them that economic life was historically contingent. The economy wrought by industrial capitalism was a new economy, and a new economy necessitated a new relationship between the state and economic life. Industrial capitalism, the progressives argued, required continuous supervision, investigation, and regulation. The new guarantor of American economic progress was to be the visible hand of an administrative state, and the duties of administration would regularly require overriding individuals' rights in the name of the economic common good.

The progressives' break with their classically liberal roots was one of the most striking intellectual changes of the late nineteenth century, one with far-reaching consequences. Progressives embraced holism, drawn by a powerful confluence of postbellum intellectual currents: the German Historical School's view that a nation was an organism, something greater than the sum of the individuals it comprised, Darwinian evolution's implication that the individual's inalienable natural rights were only a pleasant fiction, the Protestant social gospel's move from individual salvation to a collective project of redeeming America (indeed, the world), and the liberating effects of philosophical Pragmatism, which seemed to license most any departure from previous absolutes, provided it proved useful.

Ely, firing the early shots of his AEA insurgency, made clear that progressive economists rejected the "fictitious individualistic assumption of classical political economy" and instead placed society above the individual.[22] Washington Gladden, a charter member of the AEA and America's most influential social gospeler, condemned individual liberty an unsound basis for a democratic government. The tradition of respect for individual liberty, Gladden preached, was "a radical defect in the thinking of the average American."[23]

John R. Commons said that social progress required the individual to be controlled, liberated, and expanded by collective action.[24] Columbia progressive economist Henry R. Seager (1870–1930), an early student of Simon Nelson Patten's, declared that the industrial economy had simply obviated the creed of individualism.

The progressive economists' rejection of individualism and their embrace of what Daniel Rodgers calls the "rhetoric of the moral whole," was perhaps best embodied in Edward A. Ross's concept of *social control*, which referred broadly to all means, public and private, by which "the aggregate reacts on the aims of the individual, warping him out of his self regarding course, and drawing his feet into the highway of the common weal."[25] Individuals, Ross maintained, were but "plastic lumps of human dough," to be formed on the great "social kneading board."[26]

Ross was not merely touting bigger government. He was asserting that the autonomous, self-reliant individual, a figure in both the liberal and republican traditions, was now a fiction in the age of industrialization. An industrial society shaped and made the individual rather than the other way around. Granted this premise, the only question was who shall do the shaping and molding. The progressives concluded that the shaping and molding should be done by the best and the brightest, those who, uniquely, ignored profit and power to serve the common good—which is to say, the progressives themselves.

Ross's conception of social control owed most to the pioneering American sociologist Lester Frank Ward (1841–1913). Ward was a member of the Generation of 1840, who, like Ross, had been born into humble circumstances on the American frontier. Trained in paleobotany, Ward worked with John Wesley Powell in the US Geological Survey, an early and influential locus of government science. A polymath, Ward became the first president of the American Sociological Association (1906–1907), and he also served as an officer of the American Economic Association. Ward obtained an academic appointment only in 1906, near the end of his career, but he was the intellectual spearhead of the progressive assault on laissez-faire.

Ross married Ward's niece, Rosamond, and referred to Ward as "my Master." He dedicated *Social Control* to Uncle Lester and named two of his children for him.[27] Ross once opined that if he had met Aristotle socially, he doubted he would find him "a bigger man than Lester F. Ward."[28] Some historians have been nearly as reverent, designating Ward as the father of American sociology, celebrating him as the architect of the American welfare state, and, echoing Ross, conferring upon him the title of "the American Aristotle."[29]

Ward brought to American social thought two claims, both of which became pillars of progressive thinking: first, humanity was the agent of its own destiny, and second, society, not the individual, was the proper unit of explanatory account.[30]

Ward and Ross, like all reformers, lamented the changes wrought by the industrial transformation of American life. At the same time, however, they saw opportunity in society's very mutability. The industrial revolution had demonstrated that the social order could be changed, and changed rapidly and extensively.[31] Changing the social order was clearly possible, and the task and opportunity of the reformer was to control change to ensure it was progressive.

Ward's innovation was not the idea that people can shape their own destinies. American individualism was nothing if not purposeful. Nor was it the concept of progress as such. Both purpose and progress were integral to the free market economics of Herbert Spencer, Ward's biggest target.

What Ward and his progressive successors accomplished was to fashion and legitimize a methodological holism suitable for America. Ward called it the "collective mind of society."

The progressives would give it many names: nation, state, society, the commonwealth, the public, the people, the race, and, especially, the social organism. But they always gave the moral whole primacy over the individuals it subsumed.

Two questions of enormous import loomed. What did society want? And who would be charged with knowing what it wanted?

THE WISE MINORITY AND THE MORAL WHOLE

The second question was easy for progressives. The experts were charged with knowing what society wanted. With their scientific knowledge and public virtue, the experts were uniquely placed to identify Ward's collective mind of society and to act in its best interests. The wise minority, as Ross put it, should be in the saddle.[32]

Theodore Roosevelt, for one, was unafraid to say as much. "I do not represent public opinion: I represent the public," he declared. A true leader, Roosevelt explained, recognized the vast difference between "the real interests of the public and the public's opinion of these interests." By implication, Roosevelt knew what the public's interests were.[33]

The progressives developed elaborate, often anthropomorphic depictions of society as an organism, as we shall see in Chapter 6. Henry Carter Adams said the social organism had a "conscious purpose." Political journalist Herbert Croly conceived of the American nation as "an enlarged individual." Ross described society as "a living thing, actuated, like all the higher creatures, by the instinct for self-preservation." The state, Richard T. Ely declared, was "a moral person."[34]

The social organism, like any organism, subsumes its constituent parts, and progressives routinely disdained individual liberties as archaic impediments to needed social and economic reforms. The freshly founded *The New Republic* portrayed Constitutional protection of individual liberties as quaint and retrograde. What inalienable right has the individual, its editors asked, "against the community that made him and supports him?"[35] The answer was "none."

The New Republic merely echoed what was a commonplace of progressive legal scholarship. Woodrow Wilson, when president of Princeton University, dismissed talk of the inalienable rights of the individual as "nonsense."[36] Roscoe Pound said in 1913 that the Constitution's Bill of Rights amendments, the core of American civil liberties, "were not needed in their own day, [and] they are not desired in our own."[37] Charles Beard's withering analysis of the economic origins of the Constitution depicted constitutional law as no more than a tool by which the wealthy oppress the weak.[38] The progressives' discrediting of individual rights was unprecedented, but it was consistent with their view that the health, welfare and morals of the social organism came first.[39]

Progressives' social-organism talk was partly motivated by the US Supreme Court decision that limited-liability corporations were legal persons, entitled to some of the same liberties that protected natural persons from the state.[40] The progressives had an ambivalent relationship with the consolidated firm. As we shall see in Chapter 4, the progressives admired the scientifically managed firm as model of efficiency all organizations, including government, should emulate. However, they believed that business must answer to the state, which it would have to do if the state were an even larger organism, one that subsumed corporate and natural persons alike.

It was one thing to say, as so many progressives did, that society was a person, just as the Court had said a corporation was a person. Many of the analogies held good. The constituent elements (citizens, employees) were subordinate to the purposes of the whole. A well-run society, like a well-run firm, enlisted the aid of expert administrators. Indeed, as Richard T. Ely claimed, administering a great city was a harder job than running a great railroad company.[41] But what did the social organism want? The purpose of the corporation was to maximize profit. Scientific management might find ways to increase profit. The state might tax or regulate the corporation's profit. But the corporation's goal was clear. What were the analogous purposes of the social organism? What was the public analog to corporate profits, the end to which public administrators applied their expertise?[42]

The progressives answered variously, but nearly all agreed that expert public administrators do not merely serve the common good, they also identify the common good. The expert instructed on how to achieve society's goals, and also on what society's goals should be.

3

Becoming Experts

From 1880 to 1900, American political economy reconstituted itself as an academic discipline. But the new professors of economics were not content to study economic life. They were progressives who wanted to reform it, which required finding a market for the economic expertise they now claimed to possess.

When the AEA was founded in 1885, the market was thin to nonexistent, especially in business. Of what use to the practical man in industry was a college professor of political economy, armed with abstract doctrines and social gospel pieties? Prospects in government were better, but only marginally.

The US government operated a single social welfare program, the Pension Office, which by 1900 dispensed funds to one million Civil War veterans and their survivors. A sprawling bureaucratic colossus, the Pension Office supported a vast rent-seeking industry of attorneys and examining physicians, who conspired with applicants to defraud the government. A corrupt machine that disbursed public monies to buy political support, the Pension Office was no model for the administrative state. It was, rather, a cautionary tale.[1]

The federal agencies held more promise, but it was too early to tell how much. The US Bureau of Labor Statistics (1885) had only just opened. The Interstate Commerce Commission (1887), formed to regulate the railroads, was two years off. The administrative state was coming, but its size and scope were all but invisible on the horizon.

The fledging economists, fingers in the wind, cast their lot with the administrative state, which, together with the American university, was to be the great benefactor of twentieth-century American economics. Francis Amasa Walker's presidential address to the AEA in 1888 presciently understood that an alliance with the administrative state would allow economics to be in the nation's service as well as its own.

IN THE NATION'S SERVICE AND ITS OWN

Forty-eight years old, Walker (1840–1897) was the United States' most prominent Gilded Age statistician and political economist. By 1885, when the Young Turks founding the AEA tapped him for its inaugural presidency, he was already president of the Massachusetts Institute of Technology (MIT) and the American Statistical Association. Walker had collected honorary degrees from Harvard and Yale, soon to be augmented by the same from Columbia, St. Andrews, Dublin, Halle, and Edinburgh. A much decorated Civil War officer, Walker directed the 1870 and 1880 US Censuses, served as US Commissioner of Indian Affairs, and presided over MIT for fifteen years until his untimely death in 1897.[2]

Walker's premises were epistemic and vocational. He assumed, first, that the economists knew something government did not already know. Of what use was economic expertise to government otherwise? Second, he assumed that economic expertise in the service of the administrative state would advance the professional fortunes of the new discipline.

There were two important implications. The first Walker left implied: economists had to establish that their advice was objective: "disinterested" was the term of art. Working for the public interest meant avoiding too close an association with any class or special interest, even one as important to progressive economics as labor.

The second implication Walker made clear: American economics would have to shed any remaining crust of laissez-faire dogma. Laissez-faire, of the sort that had characterized mid-century American political economy into the 1870s, was a nonstarter as a professionalizing strategy. How much scientific expertise, Louis Menand writes, was required "to repeat, in every situation, 'let the Market decide'"?[3]

Minnesota economist W. W. Folwell, addressing the AEA as acting president in 1892, granted that laissez-faire, whatever its shortcomings, had been adequate for the American economy at mid-century, when "labor" was a single hired hand or a small group of journeymen and apprentices, and "capital" was the scrimping of years of toil and self-denial. But that economic era was gone, never to return. Large-scale production, the massing of capital, the advent of labor unions, and extensive immigration had rendered laissez-faire economics obsolete. A new economy, Folwell said, required a new economics, which the AEA economists had on offer.[4]

The progressive economists had already decided that laissez-faire was morally unsound and economically obsolete. What Walker did was to hammer a third nail into the coffin. Laissez-faire was inimical to economic expertise and thus an impediment to the vocational imperatives of American economics.

Having served in government in many pioneering capacities, Walker already appreciated how economic expertise cashed out in political authority. If economists could seize the scientific authority to decide which government investigations and regulations served the public good and which did not, they would never lack for work. The tasks of the administrative state were never-ending. Moreover, when the administrative state expanded—and there was every indication (and hope) that it would—the work of economists would grow with it, which, in a virtuous vocational circle, would "heighten popular interest in political economy, increase the number of its students, and intensify the instinct of union and cooperation." Rising to his theme, Walker asked, "in such a work who would not wish to join?"[5]

Walker understood that the economists' eagerness to offer their expertise was not sufficient. The client had to demand it. Fortune intervened in the form of the hydra-headed economic crisis of the 1890s. The Panic of 1893, a classic bank run, plunged the United States into the deepest depression it had yet experienced. The unemployment rate soared to 18 percent in 1894, staying above 12 percent for five years.[6] The American frontier "closed" in 1890, and with it the traditional safety valve for urban unemployment, westward migration to claim free land.

Bank failures multiplied, the stock market plummeted, and railroads and other businesses fell into bankruptcy. The 1894 Pullman strike brought the US transportation network to a standstill until President Grover Cleveland sent in US Army troops to break the strike. In that same year, punished by deflation and unemployment, Coxey's bedraggled army of the unemployed marched to the steps of the US Capitol, demanding a hearing. The national mood was so parlous that President Cleveland risked his life by undergoing secret surgery for mouth cancer at sea aboard a rolling yacht, a voyage the White House purported to be a four-day fishing trip.[7]

Financial crisis, economic panic, violent labor conflict, a political war over monetary policy, and the takeoff of the industrial merger movement combined to generate a groundswell of support for economic reform. The prolonged crisis of the 1890s lent ever-growing credibility to Walker's idea that advising or serving in government was a surer route to professional success than was the traditional public-intellectual model of shaping public opinion with newspaper columns and Chautauqua lectures.

By the time the economy recovered from the depths of the mid-1890s depression, the professional advantages of government as a client—*the* client—for professional economic expertise were almost taken for granted. In his presidential address to the AEA in 1898, Arthur T. Hadley put the prevailing view plainly: influence in public life was "the most important application of our studies." The greatest opportunity for economists, Hadley urged, lay "not with students but with statesmen." Hadley, who became president of Yale University the following year, saw economists' brightest future not in the education of individual citizens, but in "the leadership of an organized body politic."[8]

Henry Rand Hatfield, the Chicago Business School's first dean, reported on the proceedings of the massive Trust Conference of September 1899, organized by the Chicago Civic Foundation. The gathering attracted hundreds of delegates, featured a William Jennings Bryan keynote, and speeches from state governors, organized labor leaders, railroad barons, Grangers, bankers, anarchists, traveling salesmen, newspaper editors, board of trade men, and trust lawyers—one hundred presentations in all.

The professors of economics, notably John Bates Clark, made the biggest impression. "Neither the scheming politician, nor the unbalanced enthusiast, nor the unfortunate victim of industrial changes," Hatfield wrote, could be relied on. Calm, measured, and disinterested, the economists offered the best judgment of current industrial conditions and the wisest guide to appropriate legislation.[9]

An academic accountant with a PhD in political economy, Hatfield's judgment was not unbiased. But it did show how academic experts wished to be regarded, as a disinterested policy elite, uniquely positioned above the political and commercial fray, and concerned only with a public good they could identify.

In his presidential address before the AEA in 1902, Edwin R. A. Seligman of Columbia University (correctly) reckoned that the industrial merger boom promised more extensive government regulation of the trusts. The coming transformation in the relationship between government and big business would demand economic expertise and enhance the new profession's status.

Seligman was a circumspect man, careful with his words, but he did not feel the need to mince them here. Economics, Seligman told his receptive confreres, was going to be the basis of social progress, and economists were going to be the creators of the future; indeed, they would be the philosophers of American social life. A grateful public would pay deference to the economist's expertise.[10]

Hadley's second presidential address was nearly as audacious. His first, in 1898, had made the vocational case for economics in the service of the state. His second address in 1899 made the epistemic case for economics in the service of the state. Economists, Hadley claimed, functioned as the agents of the common good, "representatives of nothing less than the whole truth."[11]

John R. Commons, who had been fired from Syracuse University and had no current academic position, was to discuss Hadley's paper. Commons unloaded both barrels, mocking Hadley's idea that economists should not represent a class interest. Taking labor's side might well be the best way, Commons shot back, of promoting the welfare of all. And no client gives a fig about the common good, Commons added. What clients wanted was an expert who will say that what is good for the client is good for all.[12]

It was a stunning outburst. One discussant had to be excused to gather himself. But Commons soon changed his mind about what economic experts do, perhaps because he became one. His 1899 work as an immigration expert for President McKinley's Industrial Commission helped Richard T. Ely secure a chair for him at Wisconsin. Or maybe Commons came to see the wisdom in Seligman's reply, which acknowledged class conflict but argued that economists should not be combatants when they could be adjudicators or reconcilers.[13] Whether it was the former or the latter, or both, Commons made a long career exemplifying the very role he briefly and memorably disparaged.

At the turn of the century, American economists had barely begun to establish themselves as professionals. Their beachhead in the universities was modest. But even at this early moment, the discipline's leaders envisioned a social role that went far beyond instructing students or nudging public opinion. If administrative government was henceforth to be the guarantor of American economic progress, then the expert economist must lead.

MARKET FAILURE AND GOVERNMENT FAILURE

Behind the progressive case for an administrative state, of course, was the view that American markets no longer served the public good. Market failure was not new to Anglophone political economy. Its leading textbook in the latter half of the nineteenth century, John Stuart Mill's (1848) *Principles of Political Economy*, explored at length the many ways in which markets go awry. Markets failed to provide valuable public goods. Markets, as in the cases of railroads and utilities, permitted monopoly. Markets imposed spillover costs, such as pollution, on third parties without their consent.[14]

Markets strangled unprotected infant industries. Markets created agency problems, as when business managers pursued private ends rather than carry out their fiduciary duties to investors. And, even when they didn't fail in these ways, markets distributed their benefits unequally or unfairly. There was nothing in capitalism, Mill made clear, that ensured a just distribution.

So, why had the progressive economists' German professors disparaged Mill as the very incarnation of "English economics," their epithet for the liberal tendency of political economy in Great Britain? Mill was no apologist for capitalism. When he wrote, "laisser-faire should be the general practice," he was not uncritically extolling the virtues of free markets, the manifold failures of which he had so scrupulously catalogued.[15] Mill, rather, feared that government cures were worse than market diseases, and he spoke from experience.

The Germans, who had little experience of and less regard for the liberal tradition, facilely equated the nineteenth-century tendency toward laissez-faire with the free-trade "Manchestertum." But the public's inclination toward laissez-faire was not founded on an uncritical belief (or indeed any belief) that Enlightenment philosophers and political economists had shown conclusively that unfettered private enterprise would ensure the greatest good for all.[16] Laissez-faire's standing derived far less from worshipful celebrations of capitalism's self-regulating powers than it did from prolonged contact with government failure.

Agency problems afflicted government bureaucrats no less than it did business bureaucrats. A career civil servant, Mill warned that government was badly informed, its employees were mediocre and often corrupt, and, moreover, politics continually threatened the reform goals of efficiency and fairness alike. Mill did not believe that government should do nothing. He merely disputed the idea that government can do anything it proposed to do.

A Millian skepticism toward government's motives and competence was well founded, and arguably mandatory, in Gilded Age America, the notorious heyday of spoils-system patronage and ward-heeling machine politics. But, nearing the turn of twentieth century, American progressive economics bore no traces of Mill's pragmatic caution. The progressive economists radiated confidence in their scientific competence and in the governments they said would deploy it.

The new economics, they claimed, could diagnose market ills and prescribe remedies that would treat or cure them. Within certain limits, Ely announced in his influential textbook, "we can have just such a kind of economic life as we wish."[17] Hadley stopped short of Ely's hyperbole, but he was no less sanguine about the future. American economics, Hadley said, was "at the height of its prosperity."[18]

Seligman's presidential address confidently predicted that the new economics, like a natural science, would understand economic forces so well it would "control them and mould them to ever higher uses."[19] As mentioned earlier, Seligman's confidence was sufficient for him to portray economists—only barely established as professionals—as the creators of the future, the philosophers of social life who deserved the deferential gratitude of the American people.

In his 1910 presidential address to the American Association for Labor Legislation (AALL), Yale's Henry Farnam captured the extraordinary self-assurance of economists when he compared scientific progress in economics to scientific progress in surgery.[20] Surgery, Farnam said, was once primitive and dangerous; it did patients more harm than good. But recent advances in medical science, especially the revolutionary discovery that germs cause infectious disease, had made surgery a positive benefit to society.

Without identifying the comparable scientific revolution in economics, Farnam told the gathered labor reformers that economics, like medical science, now possessed scientific knowledge sufficient to ensure that its reform cures were "more effective and less dangerous."[21]

It was a bold claim. Economists did not possess the kind of scientific knowledge produced by the sciences they invoked as exemplars.[22] But this inconvenience did not stop them from representing economics as an established science.

Progressives of all types tended to venerate science, even or perhaps especially when they had little contact with it. Before the First World War, American economists, like all progressives, were profoundly optimistic that scientific reform would improve the economy, government, education, philanthropy, medicine, religion, the family, even humanity itself.

Science, most clearly for the social gospel progressives at the forefront of American economic reform, was also a place of moral authority where the public spirited could find religious meaning in scientific inquiry's values of dispassionate analysis, self-sacrifice, pursuit of truth, and service to a cause greater than oneself. In the progressive conception, the scientist's motives were pure.

A commitment to disinterested truth-seeking starkly distinguished the scientist from the capitalist grubbing profit and the politician chasing power. When Ely referred to economists as "natural aristocrats," he meant that their authority derived from ownership of scientific knowledge, making them as incorruptible as the propertied citizen of republican ideology.[23] When invoking science, then, the progressive economists claimed not only scientific knowledge but also scientific virtue.

Some progressive formulations distinguished between the scientist and the applied scientist. When Edward A. Ross famously described Progressivism as "intelligent social engineering," he was idealizing the expert economist as an applied scientist.[24] The social engineer worked outside politics (or, better, above it), proceeded rationally and scientifically, and pursued neither political power nor pecuniary gain but only the public good, which the engineer could identify and enact. It was the scientific spirit, Ross said, that provided "the moral capital of the expert, the divine spark that keeps him loyal and incorruptible."[25]

The metaphor of the social engineer embodied the extravagant faith of Progressive Era economists in their own wisdom and objectivity, a mostly unquestioned assumption that they could and would represent an identifiable public good. Ross's metaphor also embedded the tantalizing claim that America's economic challenges were as comprehensible and tractable as the purely technical problems addressed by engineers on the factory floor.

The engineering metaphor turned incorrigible differences into preventable errors. Financial crisis, economic panic, violent labor conflict, and money wars were thus tamed into bad design, unthinking convention, and unscientific management. Errors can be located and fixed. In an era of ramifying crises, the social engineer was an appealing conceit.

The economists' outsized confidence in their own expertise as a reliable, even necessary, guide to the public good was matched by their extravagant faith in the transformative promise of the administrative state. On its face, this was a puzzle. Progressive economists, like all progressives, regarded American government and its party system as corrupt, wasteful, and disorganized—a travesty. Why would they place their fondest hopes for economic reform in an institution they judged wholly inadequate to the task?[26]

The answer, of course, was that progressives planned to reform government and the party system as well. During the Progressive Era, then, government served a dual role for progressives—simultaneously an instrument and an object of reform. Some, like Henry Carter Adams, said government employees were mediocre because government was weak and corrupt. Make it stronger and more public-spirited, and a better class of employee would serve. It was a theory, but right or wrong, when progressives held up government as the new and necessary agency of American economic progress, they were assuming an administrative state.

They got it, eventually, but the old political order of party bosses and mossbacked judges did not simply walk offstage so that the progressives could commence administration of economic and political life from quiet rooms. The old political order resisted the progressive restructuring, and fiercely. The reforms that built the American administrative state were not immaculately conceived; they were instead produced with the help and hindrance of a mostly unreconstructed political tradition, which, unlike the progressives, understood that politics was not bureaucracy, and bureaucracy was political, too.

THE AALL: THINK TANKERY

The AALL exemplified Progressive Era economic reform. It was unusual only in that it was organized, staffed, and led by university-trained economists and public administrators—Richard T. Ely, John R. Commons, Irving

Fisher, Henry Rodgers Seager of Columbia, and Princeton's William Willoughby. The AALL leaders were crusading activists in the cause of conserving human resources (its motto), but they claimed the mantle of objectivity. They were hostile to laissez-faire but wary of state socialism. They conceived of their work as for the benefit of labor, but they regarded their own expertise as the more reliable, even necessary, guide to the public good.

The AALL masthead mapped the interlocking directorates of American Progressivism, featuring Jane Addams; Paul Kellogg, director of the Pittsburgh Survey and editor of *The Survey*; Louis Brandeis, AALL legal counsel until appointed in 1916 to the US Supreme Court by Woodrow Wilson; and Wilson himself. When Brandeis departed, Felix Frankfurter replaced him.

The AALL was hived off its parent, the AEA, in 1905, so that its advocacy could be distanced from the AEA.[27] The AALL was a pioneer in the new social space colonized just outside the expanding boundaries of the university and government. Its raison d'être was to use scholarly methods to sway legislation, but it settled just beyond the jurisdictions of the university and the government—close enough to influence, but far enough to claim independence and to avoid the institutional constraints that bound professors and government employees.

Charting new territory that would soon be settled by the first American think tanks, the Carnegie Endowment and the Brookings Institution, the AALL, like its organizational parent, grappled with the tension between its crusading activism (what Mary Furner called advocacy) and the vocational necessities of a scientific and professional standing (what Furner called objectivity).[28]

The AALL's investigations were represented as expert and thus scientifically privileged, but they were also intended to create moral outrage, mobilize public opinion, and convince government to regulate factory safety, restrict working hours, fix minimum wages, and compensate industrial accident victims—in other words, to advocate. Advocates don't care about the truth, except as it affects their case, and experts are supposed to care only about the truth.

Advocacy threatened truth, and so too did truth threaten advocacy. As experts, AALL members claimed to represent the public good, not merely the good of labor. As cofounder Adna Weber put it, the AALL's purpose was to ensure that American labor reform was not "left mainly to the laboring classes" but was assisted and guided by the experts who knew best.[29]

The progressives' presumption infuriated Samuel Gompers of the American Federation of Labor, who, though his name was on the AALL letterhead, believed that workers were better served by organized labor bargaining on its own behalf than by government experts conducting wage arbitration, in the name of the common good. After a falling out with the AALL, Gompers gleefully referred to it as the "American Association for the Assassination of Labor Legislation."[30]

Josephine Goldmark, in her memoir of Florence Kelley, recalled the quip of a friendly visitor to New York's Charities Building in 1906: "What's this bunch calling itself today?" That day the bunch was being addressed by John R. Commons, who had traveled from Wisconsin to introduce the newly formed AALL.[31] When Commons returned to the Charities Building the following year to address the New York School of Philanthropy, he termed the AALL's amalgam of expert social science and reform activism "constructive research."[32]

Commons distinguished constructive research from academic research, which sought truth for its own sake without regard to its practical applications or to its consequences for interested parties, and from journalistic research, or muckraking, which aimed at exposure. Muckraking might diagnose economic ailments and shame those judged responsible, but it could not cure or prevent them. And purely academic research did not attend to cure or to prevention. What the American economy needed, Commons said, was diagnosis, cure, and prevention, and only economic experts could supply them all.[33]

New York City Comptroller Herman Metz, an influential member of the municipal reform movement, put Commons's argument compactly. The practical man, Metz said, "knows *how*. The scientific man knows *why*. The expert knows *how* and *why*."[34] Moreover, because science ensured objectivity, the expert remained immune from partisan and class bias.[35]

Commons's formulation of the expert as a technologist, the one who knew how and where to apply science, was likely borrowed from Thorstein Veblen. Science, Veblen wrote in 1906, "creates nothing but theories. It knows nothing of policy or utility, of better or worse." It was the technologist, quintessentially the engineer, who applied scientific k[illegible] in industry, agriculture, medicine, sanitation, and economic refo

By most measures, the AALL was very successful. It

credit for several pillars of the administrative state. It won a it secured safety regulation of factories in which workers were being poisoned and disfigured by phosphorous. Workman's compensation—mandatory insurance for industrial accidents—was another AALL coup. The minimum wage campaign of the 1910s was on a multistate roll until unexpectedly derailed by the US Supreme Court in 1923.[37] The 1915–1920 campaign for government provision of health insurance, led by Irving Fisher, did not succeed, but the AALL blueprint was put to work fifty years later.

PROGRESSIVES LEFT AND RIGHT

Historians ordinarily characterize such AALL stalwarts as Irving Fisher, Henry Farnam, and Frank Taussig as conservatives, which is adequate as a first approximation. None of them were evangelical firebrands like their colleagues Ely, Commons, and Henderson. And they were situated in the colonial era colleges, Harvard and Yale, which stood at one remove from the hubs of economic reform at Johns Hopkins, Wisconsin, Columbia, and Pennsylvania. But these men were also labor legislation activists, entirely committed to the idea, as Fisher put it, that "not only can society be reformed, but to do so is the principal service of economic and social sciences."[38] Farnam, heir to the New Haven Railroad fortune, funded the AALL out of his own pocket in its early years.

The AALL reminds us that Progressivism was first and foremost an attitude about the proper relationship of science and its bearer, the scientific expert, to the state, and of the state to the economy (and polity). University-certified experts advised or served the administrative state in the fourth branch of government, which investigated and regulated economic life in the name of an expert-identified common good. But "progressive" is a political term, and political historians tend to an ideological lens. Ideology is a useful tool of taxonomy, but when it is reduced to one dimension, it is the enemy of nuance.

Forcing the progressives to be left and their critics to be right multiplies misconceptions. Theodore Roosevelt was a conservative, but he must be counted among the progressives. Call him a right progressive, but Roosevelt was an exemplary progressive. Campaigning for president atop the Progressive Party ticket, Roosevelt ran on a platform that repudiated "the laissez-faire theory of political economy and fearlessly champion[ed] a system of increased Governmental control."[39]

Right progressives, no less than left progressives, were illiberal, glad to subordinate individual rights to their reading of the common good. American conservative thinking was never especially antistatist. On the contrary, its Hamiltonian tradition had long been identified with vigorous national government, which is precisely what Herbert Croly had in mind when he famously defined Progressivism as Hamiltonian means to Jeffersonian ends.[40]

Consider reform economics' bête noire, William Graham Sumner, the man whose influence Ely said he had organized the AEA to attack. If progressives are understood to be left, then Sumner must be right. But Sumner was no conservative; he was a classical liberal. A one-dimensional analysis elides this vital distinction.

As a liberal, Sumner believed that individuals constitute and are prior to society. The progressives, who were illiberal, believed that society comprises and is prior to the individual. The progressive conception of society as an evolved organism, to which constituent individuals owe responsibility and deference, was, arguably, the more conservative conception, in the tradition of Edmund Burke.

Sumner defended free markets, which earned him the progressives' enmity. A consistent defender of limited government, Sumner also denounced American imperialism and colonialism. He loudly criticized the Spanish-American War of 1898 as a "petty three months campaign," and served as an officer in the American Anti-Imperialism League, a group organized to oppose US annexation of the Philippines.[41]

Many progressives, most famously Roosevelt, but also Ely, Ross, Commons, and others among the progressive social scientists, were enthusiastic about American empire.[42] If advocacy of American empire was made right, then a one-dimensional analysis must place the progressives on the right, and Sumner on the left.

Sumner styled himself a defender of the "forgotten man," condemning socialism and plutocracy as equally bad forms of government. He denounced plutocracy as "the most sordid and debasing form of political energy known to us." Sumner's hostility to the influence of money in politics drove him to attack the Republican Party, accusing it in 1909 of "a conspiracy to hold power and to use it for plutocratic ends."[43]

Moreover, Sumner opposed the tariff, a daring position that put him at odds with influential Republican industrialists and nearly cost him his position at Yale. The tariff, Sumner thundered, was an elaborate system by which corporate interests "get control of legislation in order to tax their fellow-citizens for their own benefit."[44]

Sumner was an advocate for free markets, not for American business. Business loathed the economic competition of free trade, and lobbied government for protection from it, as it does to this day. When business benefited from trade restriction, such as the tariff, Sumner was their enemy.

Sumner was a leading scholarly voice opposing American imperialism, protectionism, and plutocracy. Yet because he was the bête noire of economic reform, and reform is presumed to be left, a one-dimensional analysis required that Sumner be made an archconservative. If as a first approximation we must dichotomize, it is better to say that Sumner was a liberal and the progressives, left and right, were illiberal.

BUILDING THE FOURTH BRANCH IN WISCONSIN

The progressive economists established a beachhead in the universities by 1900, and in AALL-type think tanks soon after. The university supplied their scientific *bona fides*, and the think tanks provided them a shelter for advocacy that, more or less, safeguarded their claim to scientific objectivity. Both institutions would provide essential and permanent homes for American economics.

But the big prize was government. Professors and think-tankers might move elite opinion some. But suasion went only so far. Government compulsion promised economic reform that was faster and farther reaching.

The first prototype of American administrative government was built in Wisconsin. Governor Robert La Follette, the Republican progressive, empowered the University of Wisconsin faculty and unleashed them on the state. By 1908 all the economists and one-sixth of the University's entire faculty held appointments on Wisconsin government commissions, including Charles Van Hise, the University president.[45]

Commons, whom Ely had recruited to Madison in 1904, traveled the street connecting the University and the state Capitol so regularly he wore a groove into it. But Commons and his allies got legislation that established regulatory commissions, restricted working hours, fixed minimum wages, regulated utilities, and compensated industrial accident victims.

By 1912, two books extolling the Wisconsin Idea had been published, Frederic C. Howe's *Wisconsin: An Experiment in Democracy*, and Charles McCarthy's *The Wisconsin Idea*.[46] Howe had been a student of Ely's at Johns Hopkins, receiving his PhD in 1892, the same year Ely left Baltimore to direct Wisconsin's freshly founded School of Economics, Political Science, and History.

Howe claimed that the partisans and politicians, made obsolete by university experts like Commons, had all but disappeared from the state house in Madison. The field was left to the experts, who carried scientific efficiency into every department of the commonwealth, not least the dairy industry. In Wisconsin, President Van Hise said proudly, political science had moved away from "political" and toward "science."[47]

Howe described Wisconsin as a scientific laboratory of reform. Wisconsin, he said, was already the most efficient commonwealth in the United States, a model for America just as Germany was for the world. Germany was the world's most advanced scientific state, because it had been the first to call in the experts, but Wisconsin was not far behind.

McCarthy's account proposed that German heredity was also at work. Wisconsin was a "German state," because so many Wisconsinites were of Teutonic stock, "Forty-eighters" and their descendants. *The Wisconsin Idea* breathlessly described Richard T. Ely as a pupil of German professors, who returned from Germany with German political ideals to teach a German-inspired economics at a German university (the University of Wisconsin) in the German state of Wisconsin, where the young men he most inspired were, yes, of German stock.[48] Theodore Roosevelt, who wrote a flattering introduction to McCarthy's book, was unfazed, perhaps because he was in search of votes, or perhaps because he himself had imbibed a large draft of Teutonism in his graduate student days at Columbia.

As we shall see again in Chapter 7, Germany could make certain American progressives weak in the knees. Here was a state that truly understood the value of ideas, John Dewey observed in 1915. Germany had arranged its educational and administrative agencies to ensure that scientific ideas informed practical affairs. German higher education, Dewey said, was "really, not just nominally, under the control of the state." State control ensured that it had direct access to the best ideas and a ready supply of university graduates to staff its professional civil service, the bureaucracy. Even better, Dewey noted, Germany subordinated its legislature to the bureaucracy, which conducted the real business of government—administration.[49]

Dewey was talking about Hegel, and Frederic Howe was talking about cows.[50] But both were arguing that the university was a creature of the state and should supply the state with reason and beneficial knowledge. Therein lay a crucial ambiguity, however.

The state was a multifarious thing. How exactly was science to serve it? And whom was it serving? In the more democratic conception, science served the people. It was in this spirit that the University of Wisconsin developed its extension programs, carrying scholarly and practical knowledge directly to the people.

Extension courses were one thing, but government was another. If the Wisconsin Idea was to apply intelligence to improve the lives of its citizens, this could mean giving citizens what they wanted and lacked, or it could mean giving citizens what the experts said they should want. The latter construction—uplift—appealed to many progressives like Charles Van Hise, who conceived of the public good as what was good for the public. The extremely complex problems of government should not be left to an unprepared electorate, Van Hise said; what was needed was a "government of experts."[51]

BUILDING THE FOURTH BRANCH IN WASHINGTON, DC

Before the Progressive Era (wartime excepted), American government was primarily a state and local affair. Wisconsin and other states and municipalities responded more nimbly to the American Industrial Revolution than did the US government. But the US government caught up in historically short order.

As noted in the Prologue, the "fourth branch" refers to administrative government bodies—agencies—granted broad discretionary powers to surveil, investigate, and regulate areas deemed too complex or otherwise beyond the ordinary governmental capacities of the legislative, executive, and judicial branches. Fourth branch agencies have two distinctive characteristics. First, they were chartered to be independent of their creators.[52] Second, they were, uniquely, endowed with legislative, executive, and judicial powers combined. Agencies make law—regulatory rules have the full power of federal legislation—they enforce regulation, and they adjudicate regulatory disputes.

The US administrative state was erected during Woodrow Wilson's first term and was permanently fortified shortly after American entry into the First World War. The fulcrum was the sixteenth Amendment to the US Constitution, ratified in February 1913, which made income taxation constitutional. Larger and more stable sources of federal tax revenue made the administrative state sustainable.

In 1880, the US government raised 90 percent of its tax revenue from customs duties (56 percent) and excise taxes on tobacco and alcohol (34 percent). By 1930, income taxes accounted for nearly 60 percent of US government receipts.[53]

Progressive economists, notably Edwin R. A. Seligman, played a pivotal role in laying the intellectual foundations for the US income tax.[54] Taxes, they said, were not payment for government services. Seligman argued that we pay taxes "simply because the state is a part of us." The taxpayer's duty to the state was no different than the duty to oneself and one's family.[55] By implication, taxes should vary with ability to pay. Each individual should "help the state in proportion to his ability to help himself."[56] US corporate income and inheritance taxes were soon added.[57]

The economists' advocacy for fourth branch agencies went as follows. An economic outrage occurred. A government commission was convened to find facts and make recommendations. Economists were tapped to serve as commissioners, staff members, or both. The temporary commission (usually) recommended that a permanent regulatory agency be created, ideally an independent agency staffed by economic experts with broad discretionary powers to investigate and regulate.

The first at the federal level was US Industrial Commission (USIC) of 1898–1902, chartered to figure out what the United States should do about the head-spinning centripetal aggregation of American industry. The USIC featured John R. Commons, Jeremiah Jenks of Cornell, and William Z. Ripley (1867–1941) of Harvard. Jenks' former student Dana Durand, USIC secretary, organized the experts' reports.

The USIC recommended that the US government create a permanent bureau to investigate and regulate the industrial trusts, as it already did for the railroads and banks. The US Bureau of Corporations was established in 1903, and Dana Durand was appointed to be its deputy commissioner. The Bureau of Corporations was superseded by the Federal Trade Commission in 1914.

The brutal financial Panic of 1907 led to the convening of the National Monetary Commission of 1908–1912, also known as the Aldrich Commission. Economists David Kinley of Illinois, Edwin Kemmerer of Princeton, and O.M.W. Sprague of Harvard, each of whom would become president of the AEA, all served.

The Commission's executive director, economist Harold Parker Willis of Washington and Lee University, played a key role as expert to the House Banking Committee in the passage of the Federal Reserve Act of 1913, a watershed in the formation of the American administrative state. The Federal Reserve Act created the Federal Reserve banks, institutionalized more direct government control of the money supply and commercial paper, and established closer supervision and regulation of American banking practices.

When Wilson launched the income tax, he simultaneously cut the tariff, both landmark events in the history of the administrative state. No issue was more politically perilous than the tariff, but the progressives dared to imagine a permanent agency of experts who would set tariff duties, just as the Interstate Commerce Commission set railroad rates. Wilson got a Permanent Tariff Commission in 1916, now known as the US International Trade Commission. He installed Harvard's Frank Taussig as its first chair.

These were the independent agencies, the heart of the administrative state. The Federal Reserve Board (1913) regulated money, credit, and banking. The Federal Trade Commission (1914) supervised domestic industry, and the Permanent Tariff Commission (1916) regulated international trade. Of course, there was much more.

The US government created Departments of Labor and Commerce;[58] dissolved prominent industrial combinations in the steel, oil, tobacco, and sugar industries; exempted, with the Clayton Act, labor unions and agricultural cooperatives from antitrust prosecution;[59] recognized the right of federal workers to organize;[60] mandated eight-hour work days for public-works and railroad employees;[61] mandated compensation for federal employees injured in accidents;[62] regulated the telephone, telegraph, and radio industries;[63] subsidized farmers;[64] criminalized interstate prostitution (of white women only);[65] regulated food and drugs,[66] meat,[67] and narcotics;[68] federalized western lands in the name of conservation; established a Children's Bureau;[69] prohibited trade in goods produced with child labor;[70] promoted vocational education;[71] regulated railroad rates;[72] mediated railroad labor disputes;[73] outlawed railroad price discrimination;[74] established a Shipping Board;[75] created a US Postal Savings Bank;[76] made permanent the Chinese Exclusion Act of 1882;[77] and excluded from American shores, with the Immigration Act of 1917, a meticulously enumerated compendium of racially undesirable aliens.[78]

State governments, where economic reform was more precocious, regulated working conditions, inspected factories, banned child labor, and compelled education for children. Before federal legislation superseded them in 1907, thirty-three states passed laws restricting the work hours of railroad and streetcar employees. From 1900 to 1917, thirty-seven states mandated worker's compensation, an effort spearheaded by the AALL's research and lobbying. Thirty-nine states passed "mothers' pensions" laws, payments to single mothers with dependent children, from 1911 to 1919.[79] From 1909 to 1917, nineteen states and the District of Columbia restricted women's working hours, and twenty additional states made significant increases to already existing restrictions. From 1912 to 1919, fifteen states plus the District of Columbia and Puerto Rico passed minimum wage laws for women.[80] Many state governments adopted antitrust statutes before the federal government, and many banned corporate donations to political campaigns. Local governments municipalized streetcar companies and gas and water utilities.

The bulk of the work was done by March 1917, the end of President Woodrow Wilson's first term, and the eve of the United States' entry into the First World War. The fourth branch of government was established.

* * * * *

This brief capsule is not to credit the "myth of laissez-faire," a persistent tendency in American historical writing, with roots in Progressivism itself. Progressive Era advocates of a administrative state, making their case, took rhetorical license when they portrayed the first century of the American republic as a stateless, unregulated, free-market wilderness. It was not. State and municipal governments, which had always carried out most American governmental functions, regulated an impressive array of commercial activities.[81]

First, Progressive Era economic regulation did not "bring in the state," a tired formulation that implies the American state was absent or nonexistent before 1900. The Progressive Era, rather, reconstructed the American state, transforming how (not whether) government regulated economic life. The Progressive Era permanently enlarged and strengthened government regulatory power, to be sure. But no less consequentially, it relocated regulatory power.

Early and mid-nineteenth-century economic regulation was promulgated not by bureaus staffed by experts with discretionary powers but by courts deciding legal principles and by legislatures making statutes. Judges and elected representatives, not bureaucrats, made the legal rules governing economic life.

Second, those judges and elected officials regulated economic life chiefly at the state and municipal level. Progressives wanted a national regulatory state, the better to control businesses and markets increasingly interstate in scope, and also to forestall the patchwork of state and local regulation from inducing firms to locate or incorporate in the most permissive locales.

Third, the federal government's involvement in the economy had heretofore attempted to promote business, not restrain it. The tariff, for example, was justified as protection for domestic manufacturers, and internal improvements and national banking were carried out in the name of economic development. The national administrative state, in contrast, never purported to be the friend of business. It intended to control business. The United States was the land of the trust, but, uniquely among the industrialized countries, it was also the land of antitrust.[82]

Finally, progressives justified the administrative state on grounds that judges and legislators lacked the technical expertise required to understand the increasingly complex economic matters before them. Moreover, even if judges and legislators hired experts to help them understand the matters before them, they could not be trusted to advance the public good. Scientific knowledge was necessary but not sufficient. Also needed was public virtue.

Judges were "black-letter men" who considered only legal precedent and who stubbornly upheld individual liberties, which progressives regarded as archaic impediments to urgently needed improvements in social health, welfare, and morals. As for politicians, they were too often machine operatives who served the public good only insofar as it promoted their true agenda, which was placing partisans in patronage jobs and then amassing bribes and kickbacks in exchange for government contracts, licenses, and tax and regulatory concessions.

WAR COLLECTIVISM: FORTIFYING THE ADMINISTRATIVE STATE

The United States' declaration of war on Germany early in April 1917 reinforced the administrative state in two important ways. It expanded and fortified the American fiscal state, and it enlisted for war mobilization a cadre of economic experts to measure the United States' productive potential and to direct wartime economic planning.

Large wartime expenditures and decreased tariff revenues from the wartime decline in international trade gave rise to the Revenue Act of 1917, a crucial watershed in the development of the American fiscal state. It raised federal income tax rates, steeply increased income tax progressivity (its top rate was 67%), greatly expanded the income tax base, taxed large estates up to 25%, and also massively taxed corporate profits with the goal of restraining war profiteering.[83]

Seligman rightly called it "the most gigantic fiscal enactment in history." Wartime tax legislation pushed US government tax rates, Seligman observed, to "the highwater mark thus far reached in the history of taxation."[84] Another economist noted that the tax revenue raised in a single year was enough to retire the whole of the Civil War debt.[85]

Though American involvement in the war was relatively brief (nineteen months), the wartime tax regime permanently enlarged the US government. Even after demobilization, federal spending, adjusted for inflation, was more than triple its prewar levels.[86] John A. Lapp, a pioneering Progressive Era chronicler of federal regulation, observed that until the advent of US income taxes, the ordinary citizen scarcely ever encountered the federal revenue agent, who was known only to distillers, brewers, and tobacco manufacturers. The wartime tax regime changed this permanently, bringing the federal government into "closer personal relations with the individual."[87] The wartime tax regime also, at a stroke, ended the long dominance of the tariff in the US revenue system.[88]

During the war, the US War Industries Board (WIB) introduced Americans to business management methods applied by the government to the entire economy. The WIB coordinated most government purchasing, determined the allocation of economic resources, established priorities in output, restricted the alcohol trade (a dress rehearsal for Prohibition), and fixed prices on commodities in more than sixty strategic industries.[89] Railroads were nationalized.

The chief of the WIB's Central Bureau of Planning and Statistics was economist Edwin F. Gay, dean of the Harvard Graduate School of Business Administration and former president of the Massachusetts branch of the AALL. Gay was a champion of scientific management, which he hailed as "the most important advance in industry since the introduction of the factory system and power machinery."[90]

Gay and fellow economists, notably his friend Wesley Clair Mitchell, who ultimately directed the WIB's Price Division, jumped at the chance to put their scientific management ideas into government practice—initially, by gathering and systematizing economic information.[91] Mitchell's Price Division, for example, produced an immense study of American wholesale prices, data crucial for allocating wartime production assignments from Washington.

When Grosvenor Clarkson, WIB member and historian, called the WIB an "industrial dictatorship," he exaggerated, but for the purposes of paying a compliment—namely, that the WIB established that "the whole productive and distributive machinery of America could be directed successfully from Washington." The war planning effort, in Clarkson's characterization, had converted one hundred million "combatively individualistic people into a vast cooperative effort in which the good of the unit was sacrificed to the good of the whole." In appraising the consequences of war collectivism, Clarkson volunteered that they almost made war "appear a blessing instead of a curse."[92]

The WIB's apparent success at war mobilization affirmed the progressive faith in expert administration, and it gave credence to the idea that scientific management could scale, that is, an entire economy could be managed no less efficiently than a single factory. John Dewey, for one, believed that the success of war collectivism was the most important result of the First World War.

It had demonstrated, Dewey said, that expert central planners could direct a vast economy from Washington. In but a few months, Dewey wrote, the "economists and businessmen called to the industrial front" had done more to demonstrate the practicability of economic planning than had a generation of "professional Socialists." The great success of American wartime economic planning, Dewey said, was a "revolution" in economics, impossible to ignore.[93]

After the First World War, American economists were not quite the "real philosophers of social life" Seligman had dared to imagine them as in 1902. But they had profitably seized the professional opportunity presented by the demands of war and reconstruction, expanding their new national role as expert advisors and policymakers. Wesley Clare Mitchell, reflecting in 1924, observed that the Great War had restored to economics "the vitality it had after the Napoleonic wars."[94] The First World War had been a global catastrophe in countless ways, but it proved to be a boon for American economic expertise in the service of the state.

EXPERTISE IN DEMOCRACY

One of the first accounts of Progressivism was Benjamin Parke DeWitt's 1915 volume, *The Progressive Movement*, which Richard T. Ely published in his long-running series of reform treatises, *The Citizen's Library of Economics, Politics and Sociology*. All progressives shared three common goals, DeWitt wrote: to make government less corrupt, to make government more democratic, and to give government a far bigger role in the economy.[95]

Granting DeWitt's characterization, significant tensions between all three of Progressivism's goals were evident. In fact, the realization of any one progressive goal worked to undermine the other two.

Progressive anticorruption reforms, such as voter registration, reduced democratic participation, often by design. Indeed, the legal and extralegal disenfranchisement of black Americans in the Jim Crow South was regularly justified as an anticorruption measure. In 1915, American participatory democracy was already in steep decline.

Textbooks remember Progressivism for its prodemocratic reforms, of which there were many. Amending the US Constitution to give women the vote and to require direct election of US Senators are celebrated examples. States also added prodemocratic measures, such as the referendum, the recall, and election of judges. Between 1903 and 1908, twelve states regulated lobbying, twenty-two states banned corporate campaign contributions, and thirty-one states mandated direct primaries.[96]

But the uplift of women had already been met with the exclusion of African Americans. In the polity no less than in the economy, progressives offered uplift to those they regarded as deserving victims and restraint to those they regarded as underserving threats. The net effect was a radical and permanent decline in American participatory democracy.[97]

Voting turnout between the 1890s and 1920 plummeted in the South, where Jim Crow legislation, abetted by a bloody campaign of racial terror, effectively disenfranchised African American men. Overall voter turnout in the former states of the Confederacy fell to 30 percent by 1904.

Many progressives turned away from the plight of black Americans, but others justified the brutal reestablishment of white supremacy in the Jim Crow South. Professor Woodrow Wilson told his *Atlantic Monthly* readers that the freed slaves and their descendants were unprepared for freedom. The freedmen were "unpracticed in liberty, unschooled in self control, never sobered by the discipline of self support, never established in any habit of prudence ... insolent and aggressive, sick of work, [and] covetous of pleasure." Jim Crow was needed, Wilson said, because without it, black Americans "were a danger to themselves as well as to those whom they had once served." When President Wilson arrived in Washington, his administration resegregated the federal government, hounding from office large numbers of black federal employees.[98]

John R. Commons approved. Black suffrage, Commons said, was not an expansion of democracy but a corruption of it. Blacks were unprepared for the ballot, and giving it to them had served only the interests of the oligarchy. Apparently forgetting the valor of the black soldiers who served in the Civil War, Commons wrote, "by the cataclysm of a war in which it took no part, this race, after many thousand years of savagery, was suddenly let loose into the liberty of citizenship, and the electoral suffrage."[99]

Edward A. Ross was not to be outdone when it came to contempt for his imagined inferiors. Black suffrage, he said, was the taproot of American political corruption. "One man, one vote" Ross wrote, "does not make Sambo equal to Socrates."[100]

Disenfranchising Southern blacks was profoundly undemocratic, but it was also, as Eric Foner observed, a typical progressive reform. The progressive goal was to *improve* the electorate, not necessarily to expand it. Indeed in Mississippi, the disenfranchisement of black citizens was immediately followed by progressive measures establishing direct election of judges, the ballot [illegible]e, and the referendum.[101] By choosing quality over quantity, the [illegible]ent, democratic deliberation among the remaining voters was

The Progressive Era decline in participatory democracy could not be laid solely at the feet of Jim Crow. Voter turnout dropped everywhere. In New York State, turnout fell from 88 percent in 1900 to 55 percent in 1920. In national elections, turnout dropped from 80 percent in 1896 to less than 50 percent in 1924.[102] Measured by voting rates, there was less democracy, not more.

A second tension in DeWitt's troika of progressive goals was less conspicuous but just as fundamental: the tension between economic expertise and democracy. Economic progressives certainly wanted to give government a far bigger role in the economy, but by means of an administrative state. How could progressives return government to the people while simultaneously placing it beyond their reach in the hands of experts?[103] They could not.

The very rationale for administrative government was to separate administration from politics, to move decisionmaking power away from voters and their elected officials (and judges) and vest it with the scientific experts in the independent agencies, such as the Federal Reserve Board, the Federal Trade Commission, and the Permanent Tariff Commission. If democracy meant, as DeWitt characterized it, control of the many, then government by expert administrators was, by its nature (and indeed, by design) less democratic.

The dilemma was unavoidable, and it remains so today. Democracies need to be democratic, but they also need to function, and nearly all progressives believed that the new industrial economy necessitated a vigorous administrative state guided by experts.

Economic reformers fell into two camps regarding the tension between expertise and democracy. The more egalitarian progressives, such as Jane Addams and John Dewey, wanted more democracy and more expertise, but never really figured out how to get both. How could progressives redeem the democratic system by an undemocratic substitution of their own judgment for that of the people?[104] The more egalitarian progressives usually appealed to some notion of instruction, of the people or of their elected representatives. The universities' extension programs, bringing knowledge directly to the people, exemplified this impulse.

In Lester Frank Ward's early imaginings, the experts provided advanced instruction to the people's representatives, turning every legislature into "a polytechnic school," a laboratory of research into the laws of society. The way to ensure progressive legislation was to transform every legislator into a progressive social scientist.[105] In Ward's scheme of 1883, the expert served not as a policymaker but as the shaper and educator of the policymaker.

Ward himself did not become an academic until twenty-three years later, when the ascendant Wisconsin Idea asserted that the experts don't instruct the politicians so much as replace them. Even if Ward's fanciful vision could have been realized, his progressive successors would have deemed it inadequate. The instructed politician was not sufficient to realize the public good, because the politician lacked the public virtue of the expert.

The more egalitarian progressives wanted to believe that expertise, done right, could promote democracy rather than substitute for it. They hoped that putting scientific information in the hands of the voters would lead the electorate to make better choices and become more actively engaged in civic life. But the people invariably disappointed them. Some, like Walter Lippmann, abandoned as impossible the dream of an instructed electorate and unabashedly advocated for technocracy.[106]

* * * * *

The progressive economists—or certainly the most outspoken among them—were not egalitarians and never entertained the notion that expertise could work through the people. They were frank elitists who applauded the Progressive Era drop in voter participation and openly advocated voter quality over voter quantity. Fewer voters among the lower classes was not a cost, it was a benefit of reform.

Senator Albert Beveridge of Indiana, a leading progressive legislator and chair of the 1912 Progressive Party convention, put it baldly. The rule that just government derives its authority from the consent of the governed, Beveridge said, "applies only to those who are capable of self-government." So long as the United States was plagued with inferior races and classes, Commons said, it could not be a democracy at all, only an oligarchy disguised as one.[107]

Public education was a fine thing, Ely granted. But economic reform required the leadership and guidance of the "superior classes." Ross said self-government was feasible when the United States was an empty continent, but no more. The new economic order required the leadership of "superior men." Ross, never unwilling to be blunt, said that removing control from the ordinary citizen and handing it to the expert provided "the intelligent, far-sighted and public-spirited" a longer lever to with which to work.[108]

For the Wisconsin men, the expert was not a mere civil servant, retained by the public for the same reason it might hire a dentist or a plumber. Ely's and Ross's "superior men" were indeed superior men, members of a natural aristocracy. It was high time, Ely said, to abandon the outmoded eighteenth-century doctrine that all men were equal as a false and pernicious doctrine. The wiser and stronger were obligated to lead the feebler members of the community. The ordinary wage earner, Ely said, clearly felt the need for superior leadership.[109]

Ely granted that public education could uplift ordinary people. At the same time, he doubted that all Americans were educable. Even after instruction, some remained unworthy of the ballot. How many? Governing New York City would be easier, Ely once ventured, "if thirteen per centum of the poorest and most dependent voters were disenfranchised."[110]

Ely's elitism did not soften. It hardened. The "human rubbish heap," he wrote in 1922, was far larger than "a submerged tenth." The intelligence testers had scientifically demonstrated that 22 percent of US Army recruits were inferior.[111]

The forthright elitism of Ely, Ross, and Commons was hardly unknown in American life. But their case for public leadership by social science experts gave elitism a new form and rationale in the Progressive Era, one expanded on by Irving Fisher. The United States had abandoned laissez-faire, Fisher said, out of recognition that "the world consists of two classes—the educated and the ignorant—and it is essential for progress that the former should be allowed to dominate the latter." When America admitted that it was right for the educated to give instruction to the ignorant, it opened "an almost boundless vista for possible human betterment."[112]

The expert bettered society by regulating big business; protecting labor; and also by restraining drinking, gambling, prostitution, and indecent literature. Laissez-faire's mistake was to confuse a person's desires with what is intrinsically desirable, an error that experts overcame by giving people not what they want but what they should want.

Like most progressives, Fisher disavowed socialism, warning that when one class attempted to rule another, the result was corruption, inefficiency, lack of adaptability, and abuse of power. Having just advocated expert rule of the ignorant, Fisher simply did not consider that the experts, his class, might also fall prey to corruption, inefficiency, inflexibility, or abuse of power.[113]

In the progressive self-conception, the experts were not partisans of a class or any other interest. They were defenders of the public good, selfless scientists motivated solely by truth. H. L. Mencken once said of Theodore Roosevelt, "he didn't believe in democracy, he believed simply in government." The same sentiment justly applied to progressives like Ely, Commons, Ross, and Fisher.[114]

Progressives who did believe in democracy did not exhibit the same contempt for ordinary people. But they had to recognize the necessary elitism of expertise and the tension it created for a democracy. Edwin R. A. Seligman was alert to it. The expert economist, Seligman said, was like a priest, with priestly functions. He first instructed the people on what they should want and only then acquainted them "with the means of their satisfaction."[115] A student of history, Seligman observed that economics had been least influential in the most democratic places.[116] Unlike his inegalitarian colleagues, Seligman did not applaud the undemocratic nature of expertise. But he could not deny the fact of it. It was regrettable, Seligman concluded, but necessary.

4

Efficiency in Business and Public Administration

During the decade from 1908 to the US entry into the First World War in 1917, the economic progressives' diagnosis of the American economy was economic injustice, waste, and conflict. In a word, their cure was "efficiency." As historian Samuel Haber wrote of the era, "*efficiency* and *good* came closer to meaning the same thing than in any other period of American history."[1]

"Efficiency," like "progressive," was a term of praise. It connoted not only reducing waste to increase output, but also modernity, organization, orderliness, and objectivity. When Jane Addams argued labor legislation was necessary for "efficient citizenship," and economist Helen Sumner maintained that wage-earning women endangered "efficient motherhood," they revealed efficiency's expansive associations.[2]

ECONOMY OF SCALE

The late Progressive Era embrace of efficiency had origins in both the labor question and the trust question. During the great industrial merger decade of 1895–1904, 1,800 major industrial firms were consolidated into 170 giant firms, and nearly half of the consolidated corporations enjoyed market shares of more than 70 percent.[3] Was the consolidated industrial corporation more efficient? The progressive economists answered with a decisive "yes." As we shall see, they also advocated making government more efficient by importing modern business management practices.

Some historians have seen the progressives as inconsistent, simultaneously criticizing business-made chaos while scheming to "reorganize government along business lines."[4] The progressive economists, right or wrong, saw no inconsistency. In distinguishing the firm from the market, they were distinguishing administrative planning from decentralized market exchange.

A large firm is a bureaucratic organization. Inside it, choices are determined not by price but by command. The efficient firm was one whose choices were made by expert managers applying scientific methods. These techniques, moreover, could be applied to nonbusiness forms of organization.

In contrast, a competitive market was not a single organization but a decentralized network of many small firms and customers. Such a market might be regulated, as were the physical marketplaces of early nineteenth-century America, but it could not be administered. Market choices are decentralized, guided by prices, not by command. Market outcomes are unplanned, and *this,* many progressives argued, was the source of economic disorder and waste.

Progressive economists attacked the free market system, but they did not oppose greater industrial scale. On the contrary, they regarded the new consolidated giants as exemplary "islands of conscious power in [an] ocean of unconscious cooperation," which, unlike the small merchants and producers they were displacing, were more likely to be scientifically managed and cost efficient.[5]

Expertise lay at the heart of the progressive conception of business efficiency. Efficiency didn't arise spontaneously with growth in the size of a business. Efficiency required expert management. Indeed, large-scale enterprise became economically viable only when the visible hand of expert management proved to be more efficient than the invisible hand of market coordination.[6]

Columbia University's Wesley Clair Mitchell (1874–1948), a student of Thorstein Veblen's, and later founder of the National Bureau of Economic Research, made this distinction plain in his 1913 magnum opus, *Business Cycles.* Coordination within a firm came from careful planning by experts, whereas coordination among independent firms was not planned but was spontaneously ordered. Expert administration, Mitchell said, yielded efficiency, whereas market coordination among firms created waste. The new industrial giants were efficient precisely because they increased the scope of expert management, while eliminating the waste of market "higgling." In Mitchell's formulation, economic waste was not business made; it was market made.

Progressives regarded big business as a permanent feature of the new economic landscape. It was useless to abuse and attack the trusts, John R. Commons told a *New York Times* reporter covering the Chicago trust conference in 1899. The trusts must be discussed "from the viewpoint of inevitability."[7] Americans must recognize, said Princeton's William F. Willoughby, that industrial consolidation was not only inevitable; it was also desirable.[8]

Progressive economists argued that greater size increased efficiency in three ways. Firms merged by vertical integration eliminated market-made waste. First, no cost-increasing transactions with middlemen were required if Carnegie Steel mined its own coal and iron ore, and transported raw material to its mills using its own barges and rail cars. Second, larger industrial scale and access to lower-cost financing, provided factory workers with technically superior capital equipment, which boosted labor productivity, lowering production costs significantly. Third, in the case of industries with high fixed costs, exemplified by the railroads, horizontal mergers helped eliminate "ruinous" or "cutthroat" competition, where competition pushed prices too low to recover fixed costs, crushing wages and profits alike.

What was more, the consolidation of industry promised to reduce business cycle volatility, offering more stable employment. With unified management of an industry, overproduction was avoided, and with it the unnecessary cycle of boom and bust. William Willoughby, future president of the AALL, argued that the consolidated firm was good for workers. It offered better conditions, greater safety, steadier employment, and higher wages.

Progressive economists certainly were not apologists for big business. They condemned monopoly, which for them meant market power sufficient to charge prices above cost. Like most progressives, they feared the potential of the trusts to corrupt politics. And some, like John Bates Clark, believed that less competition in industry inhibited technological innovation. But progressives distinguished monopoly from size, and because of this, were not antimonopoly in the populist sense of the term.

Indeed, the 1895–1904 decade of industrial consolidation goes some way to explaining the puzzle of why, in 1905, William A. White could say it was " funny how we have all found the octopus," when, as Daniel Rodgers puts it, less than a decade earlier "his like had denied that animal's very existence."[9] The consolidated industrial firm "discovered" by economic reformers circa 1905 was a somewhat different beast than the railroad octopus of a generation before.

One difference was scale. The market values of the new behemoths, exemplified by US Steel's initial capitalization of $1.4 billion in 1901, were 1,000 times larger than the largest manufacturing enterprises of the 1870s. Another difference was scope. The railroad octopus squeezed western farmers and lumbermen. The new octopus, whether created by consolidation (oil, steel, coal, copper, meatpacking, sugar, tobacco) or by government monopoly (ice, subway, gas and electricity, streetcars), threatened a larger and more powerful group of consumers and firms.

In the 1880s, agrarian populists portrayed the railroad octopus with its tentacles crushing its customers as well as its competitors, the shipping and stage lines.[10] The populists' antimonopolism, continuing the Jacksonian tradition, opposed big business not only for gouging its customers but also for destroying its competition, a small-is-good position that persisted in American antitrust law well into the middle of the twentieth century. For the populists, size *was* monopoly. Progressives, in contrast, regarded small business as inefficient and outmoded, and they largely applauded its destruction.

On the question of trust policy, all three major presidential candidates in the 1912 election offered impeccable reform credentials. The Republican Taft administration had broken up Standard Oil and American Tobacco; indeed, Taft had initiated more antitrust proceedings than had Theodore Roosevelt, the "trustbuster" who was in 1912 heading the Progressive Party ticket. Woodrow Wilson, the New Jersey governor and Democratic Party nominee, was also vigorously antitrust.

There were differences among progressives concerning trust policy. The leading strand of progressive business regulation, represented by Roosevelt, argued that the industrial behemoths should be regulated but not dismantled. Rooseveltian progressives imagined the federal government as a powerful, neutral defender of the public interest in securing the lower costs provided by large scale, with vigorous oversight to ensure that the trusts did not abuse their pricing power. The aim of Rooseveltian antitrust was not to punish large size but to punish bad behavior.

Rooseveltian progressives said the dismantling of big firms was impractical and would destroy the cost efficiencies that large scale provided. Richard T. Ely argued that the consolidated giants were "a good thing, and it is a bad thing to break them up; from efforts of this kind, no good has come to the American people."[11] Willoughby said the small businesses driven to the wall were the least fit to survive.[12]

Progressive political journalists used even blunter language. Walter Lippmann, writing in *Drift and Mastery,* disparaged antitrust as destructive and retrograde:

> If the anti-trust people . . . [did] what they propose, they would be engaged in one of the most destructive agitations that America has known. They would be breaking up the beginning of collective organization, thwarting the possibility of cooperation, and insisting upon submitting industry to the wasteful, the planless scramble of little profiteers.[13]

Lippmann's *New Republic* colleague, Herbert Croly, said the small businessman should be "allowed to drown."[14]

The barons of big business found such rhetoric quite congenial. They used the language of anti-individualism and efficiency to portray industrial combination as cooperation, not conspiracy. Volney W. Foster, a man made wealthy by the Chicago asphalt trust, rendered the idea tersely: "concentration means organization, organization means system, and *system* is the great *economist.*" Combination not only reduced waste, it morally elevated industry by ending the "debauch of competition."[15]

A less obscure consolidator, John D. Rockefeller, said combination created "modern economic administration." The day of the combination was here to stay, Rockefeller proclaimed, and "individualism has gone never to return."[16] Few seemed to mind that the argument was a *non sequitur*. Even granting the premise—that market competition was not healthy rivalry to serve consumers, but, rather, an antiquated, unplanned, individualistic, and wasteful race to the bottom—it did not follow that consolidation was therefore modern, organized, cooperative, and efficient.

Louis Brandeis, who had the ear of President Woodrow Wilson until appointed to the US Supreme Court in 1916, was the rare progressive skeptical of big business. Brandeis doubted large scale was more efficient. The swollen industrial giants were not the fittest survivors of a fair economic competition. They had obtained their position through "artificial and illegitimate means of preventing competition."[17] Brandeis also warned that the Rooseveltian approach might lead to corporate capture of government regulators, enabling rather than preventing corruption.

Brandeis's battle against bigness made him an outlier among economic reformers. Harold Laski called Brandeis a "romantic anachronism."[18] Brandeis's lifelong defense of small business, which showed little concern for the even smaller consumers obliged to pay its higher prices, appeared retrograde to most progressives. Brandeis styled himself as "the people's lawyer," but, as historian Thomas McCraw observed, he was better described as the small-businessman's lawyer.[19]

Few economic reformers accepted Brandeis' claim that big business got big only through illegitimate means. Fewer still acknowledged his warning that progressives had uncritically placed their faith in the ongoing virtue and wisdom of the administrative state. Nearly all economic reformers were, like Herbert Croly, supremely confident that Hamiltonian means could be made to serve Jeffersonian ends, so long as the "wise minority" was in the saddle.

TAYLORISM

The bible of the 1910s gospel of efficiency was Frederick Winslow Taylor's (1856–1915) international bestseller, *The Principles of Scientific Management* (1911).[20] A century later, scientific management, or *Taylorism,* serves as a term of abuse. Taylorism is today most often associated with dehumanizing work practices, time and motion studies, a preoccupation with worker malingering, and the deskilling of labor. The Taylor system, on this reading, treated workers as mere cogs in the industrial machine.

Progressive economists and their reform allies regarded scientific management altogether differently. Taylor's program appealed to a great many progressives, who saw in scientific management a method for improving workers' jobs and wages, and in Taylorism a system for making factory work and other forms of organization more efficient. Taylor's biographer rightly judged *The Principles of Scientific Management* a progressive manifesto.[21]

Taylor's champion was Louis Brandeis, who called Taylor a genius and made Taylor's national reputation when he used scientific management theory to criticize the railroads in the *Eastern Rate* case of 1910. Brandeis, who represented the shippers opposed to the rate increase the eastern railroads sought from the Interstate Commerce Commission, invoked Taylor to argue that railroads would not need higher rates if only they would manage their costs more efficiently, using the principles of scientific management. Brandeis's star witness was efficiency expert Harrington Emerson, who regarded Taylor's "Shop Management" as one of "the most important papers ever published in the United States."[22] Emerson testified that the eastern railroads were wasting about $22 million daily (in 2014 dollars).[23]

John R. Commons called scientific management "the most productive invention in the history of modern industry."[24] Commons would later claim, after leading a platoon of Wisconsin graduate students through a study of thirty industrial firms, that industrial capitalism could survive, but only with installation of expert management.[25] Theodore Roosevelt saw the efficiency gains from scientific management as a vital example of national conservation. We couldn't ask more from a patriotic motive, Roosevelt said, "than Scientific Management gives from a selfish one."[26]

Muckraking journalists, who made their living treating business claims with suspicion, piled on the Taylor bandwagon with alacrity. Ida Tarbell, best known for her damning investigation of Standard Oil, described Taylor as a creative genius. No man in history, she told her readers, had done more for "juster human relations."[27] *The American Magazine*, cofounded by Tarbell, Lincoln Steffens, and Ray Stannard Baker, serialized *Principles of Scientific Management*, promoting Taylor's new science as "The Gospel of Efficiency."[28] Political journalists also embraced Taylorism. *The New Republic's* Walter Lippmann told his readers that scientific management would "humanize work."[29]

Florence Kelley, like many progressive leaders, joined the Taylor Society, which, during the 1920s, served as refuge for future New Dealers, such as Rexford Tugwell and John Maurice Clark. Tugwell, one of Franklin D. Roosevelt's "Brains Trust," would later say, "the greatest economic event of the nineteenth century occurred when Frederick W. Taylor first held a stop watch on the movements of a group of shovellers in the plant of the Midvale Steel Company."[30]

Thorstein Veblen was once skeptical of scientific management, but he became one of its most zealous converts. Emboldened by the apparent success of American economic planning during the First World War and optimistic about the Bolshevik revolution, Veblen proposed that the whole of US manufacturing be placed in the hands of an elite corps of experts who would be born, bred, and trained to scientifically manage American industry. The engineers of this industrial "General Staff," freed from the dictatorial rule of the "Captains of Finance," would immediately eliminate the shameful waste of profit-seeking capitalism. It was an open secret, Veblen asserted, that the moment the engineers ended capitalist mismanagement, US industrial output would grow explosively, increasing by 300 to 1,200 percent.[31]

His prophesy of class-conscious technocrats seizing control of American industry for the good of all, was, as ever with Veblen, over the top. But the essentials of his vision were widely shared among progressives. Industrial gigantism was inevitable and desirable. Efficiency came from expert management, not from market discipline. Capitalists and financiers were selfish profit seekers, but the engineers, whom Veblen elevated to Platonic Guardians, were selfless guarantors of the public good.

Veblen was an unsurpassed chronicler of human foibles, including his own. But he simply did not consider the prospect that a Soviet of engineers might fail to be selfless servants of the public good. When it came to the experts, Veblen's critical faculties abandoned him, a telling measure of expertise's sway over Progressive Era economic reform.

Progressives showered superlatives on Taylor, because he offered them an irresistible package—efficiency, labor peace, and higher wages, all realized by means of the expert application of science. First, science would improve workplace efficiency. Rather than follow arbitrary rules of thumb, the industrial engineer would, with observation and experiment, determine optimal work techniques, discovering the "one best rule." It was, Taylor said, a "science of shoveling."[32] The optimal shovel load was 21 pounds, Taylor determined, so shovel size should be adjusted for the different densities of materials a laborer carried.

Scientific management also promised to advance workplace fairness. When Taylor substituted expert planning for what he saw as the arbitrary commands of shop floor bosses, progressives hailed a new "leadership of the competent." Herbert Croly contended that scientific management replaced "robber barons" with "industrial statesman." [33] The efficient organization, Croly said plainly, "puts the collective power of the group at the service of its ablest members."[34]

Taylor claimed that his system came first, but he was not eliminating authority. He was merely relocating it. Taylor did indeed reduce the power of shop floor bosses, but he did so by giving it to himself and the other efficiency gurus.

Taylor promised labor peace by a two-step process. The first step was to establish efficient work techniques, enforced and calibrated by enhanced surveillance of worker effort. The second step was the explosive increase in productivity that ensued. The massive gains in industrial output would raise wages and profits alike. Scientific management, Taylor promised, would make it is possible for workers and management to "take their eyes off the division of the surplus until this surplus becomes so large that it is unnecessary to quarrel over how it shall be divided."[35]

In practice, Taylor rarely made it to step two. Production workers resisted being sped up, with its presumption of worker malingering. They resented the loss of autonomy and the devaluing of their craft skills. When the Taylor system was installed at the Watertown Arsenal in Massachusetts, workers staged a walkout and successfully petitioned the War Department for its removal.[36]

Taylor's governing premise—that more surveillance and less autonomy would be welcomed by grateful workers, if only the new authorities were engineers rather than bosses—was, seen in retrospect, preposterous. But, however much Taylor's claims read as hyperbole today, scientific management in the 1910s seduced progressives with its promise of a scientific solution to the labor question. Scientific management would increase efficiency, boost wages, and see that industry was governed not by self-seeking capitalists but by public-spirited experts.

SCIENTIFIC MANAGEMENT OF HUMANKIND

Progressives enthusiastically seized on industrial efficiency as an exemplar, imagining that scientific management could increase efficiency not just on the factory floor but also in all corners of an industrial society plagued by waste, conflict, disunity, and injustice. Following Brandeis's 1910 intervention on behalf of Taylor, a flood of reform volumes on efficiency appeared, preaching greater efficiency in government, in charity, in education, in medicine, in religion, in the home, and in human beings themselves. The times, said progressive sociologist Charles Horton Cooley, demanded nothing less than a "comprehensive 'scientific management' of mankind."[37]

The idea of using business methods to improve corrupt and inefficient government enjoyed great currency among political reformers. Many American cities established efficiency bureaus, spearheaded by New York City's Bureau of Municipal Research, which was cofounded in 1906 by Edwin R. A. Seligman to promote, as its motto read, "the application of scientific principles to Government," also called "efficient citizenship."[38]

Milwaukee's city government founded its Bureau of Economy and Efficiency in 1910 and tapped the ubiquitous John R. Commons to run it. Carl Sandburg, covering efficiency for *La Follette's Weekly*, ingenuously described Commons "as one of those restless, persistent geniuses of toil," whose work combating waste in Milwaukee was "blazing a way out of the civic wilderness."[39]

Municipal research, like scientific management, concerned itself with impersonal "methods rather than with men," so its techniques could be employed in any city.[40]

Other cities hired the Bureau of Municipal Research to reorganize their budgeting methods, operations, accounting, and collections. It was now a consultancy, pushing still further the boundaries of the new social space progressive scholars were colonizing—activist but scientifically expert, academic but free of university oversight, privately funded but with quasi-public authority, publicly oriented but available for hire. The next step for municipal reform was to reproduce itself.

In 1911, the Bureau of Municipal Research opened its Training School for Public Service, the first American institution dedicated entirely to training civil servants for public administration. It recruited Taylor for lectures and required all its students to read his *Principles of Scientific Management*.[41] Commons lacked the funds for a school of public administration, but he brought to Milwaukee nationally known management gurus, men like Harrington Emerson, Brandeis's star witness in the *Eastern Rate* case.

The reach of municipal reform expanded. Some American cities replaced mayors with city managers, who promised administration, not politics. A vital reform innovation placed exclusive budget authority with the city manager, leaving the elected city council members with only the choice to vote up or down on the budget they were presented.

Cities, municipal reformers said, could be scientifically administered, provided partisanship and politics were successfully pushed to one side, as Frederic Howe claimed the state of Wisconsin had accomplished by 1912. Howe, who portrayed Germany as the world's most advanced scientific state, just as Wisconsin was America's, reminds us that the progressives' gospel of government efficiency was German inflected.[42]

State governments also founded efficiency bureaus. When the efficiency contagion reached Washington, President William Howard Taft formed the US Commission on Economy and Efficiency. Taft turned to the established experts, importing the Bureau of Municipal Research's Frederick D. Cleveland to bring order to the national agencies. He also tapped William F. Willoughby, past president of the AALL. When Willoughby completed his service on the Taft Commission, he decamped for New Jersey, taking over the McCormick Chair in Jurisprudence when Woodrow Wilson left the Princeton presidency for greener pastures.

While at Princeton, Willoughby was called to direct the Institute for Government Research, a key outpost of the new social territory progressive scholars were colonizing, this time in Washington. The Institute was a think tank founded in 1916, later named for its patron, Robert Brookings. Like the Bureau of Municipal Research it emulated, the Brookings Institution conducted scholarly research on the science of administration, lobbied for a more efficient public sector, eventually founded a school for training students, and hired out its experts to government and other clients. Brookings' fortunes rose rapidly when it secured a dream client, the War Department, which was rapidly mobilizing for the First World War, flush with funds produced by a mammoth income tax increase and in need of immediate assistance.[43]

Professor Woodrow Wilson was a theorist of public administration. Subsequently, as governor of New Jersey and president of the United States, he got to put his theory into practice.

William F. Willoughby did it the other way around. Before arriving at Princeton to teach administration, he had practiced in Puerto Rico, serving as treasurer, secretary of state, and acting governor, among other posts, between 1901 and 1909. Like a number of progressive social scientists, Willoughby cut his administrative teeth by practicing on the American colonial possessions recently won from Spain. Theodore Roosevelt called Willoughby the "King-pin of Porto [sic] Rico," and it was Willoughby's years of administrative experience there that led Taft to summon him to Washington to see whether his methods could make the US government more efficient.[44]

The free hand Willoughby had enjoyed in administering Puerto Rico clearly informed his views of efficient governance. Now that American government had entered nearly "every field of activity," Willoughby wrote, popular government was no longer up to the job. What the times required, he said, was an administrative state—that is, government that employed the "same standards of efficiency and honesty which are exacted in the general business world."[45]

Administrative government, in turn, demanded that the divided powers of the US government be consolidated. The Constitution's separation-of-powers doctrine, which decentralized power by design, was as inefficient and obsolete as was the "planless scramble of little profiteers" in the age of consolidated industry. Efficient administration required that government, like industry, be consolidated, centralized, organized, and administered.

Willoughby's disregard for constitutional checks and balances was widely shared among progressive political scientists. Columbia's Charles Beard, who in 1915 was made director of the Bureau of Municipal Research's Training School, disparaged the separation of government powers as "the political science of negation."[46] Woodrow Wilson, on the campaign trail in 1912, told voters that it was time for the federal government to be liberated from its outmoded eighteenth-century scheme of checks and balances.

Government, Wilson said, was a living organism, "accountable to Darwin, not to Newton." Since no living thing can survive when its organs work against one another, a government must be free to adapt to its times, or else it will perish.[47] The adaptation Wilson had in mind was to neutralize Congress and consolidate power in a vigorous executive. It was a plan he had been elaborating and revising for more than thirty years, an early version of which, *Congressional Government*, he completed while still a graduate student at Johns Hopkins.[48]

The US government, Wilson said, was weak and slow, because its powers were divided; and it was inefficient, because it lacked the leadership of a commanding executive.[49] The president, as the only government official who faced a national election, should be "at liberty, both in law and in conscience, to be as big a man as he can," Wilson wrote. Only "his capacity will set the limit" to his power.[50]

As a young professor, Wilson did not possess the fully elaborated discourse of scientific management that flourished during his two terms as US president. But efficiency had always been at the heart of his plan to centralize government authority in a powerful executive and to professionalize the bureaucracy that would serve it. Rightly conceived, Wilson wrote, professional government administration was "a field of business," wholly removed from "the hurry and strife of politics." The public administrator *administered*; his role was no more political than "the methods of the counting-house [were] part of the life of society."[51]

The efficiency movement also made deep inroads into the philanthropy business. Scientific charity groups, such as the Charity Organization Society of the City of New York, made it their mission to bring order and efficiency to poor relief. Instead of heedless, uncoordinated dispensing of alms, organized charity promised to gather the city's diverse and duplicative charitable efforts under a single umbrella. Once consolidated, there was no possibility of forum shopping or double dipping. The charity experts could administrate—

that is, determine what kind of relief, if any, was appropriate—and direct the worthy applicant to the appropriate agency. It was, wrote Robert Hunter, social worker and author of the influential book *Poverty*, the philanthropic equivalent of an efficiently run department store.[52]

The Russell Sage Foundation was chartered in 1907 with the quintessentially progressive mission of making charity more efficient. It immediately announced that the Foundation would not waste its millions providing relief to the needy. Instead it intended to fund research into poverty's root causes.[53] Prevention was more efficient than treatment, and few reformers doubted that expert investigators would discover poverty's root causes and eliminate them.

With the cooperation of school superintendents, efficiency experts also moved into city schools. Funded by Russell Sage, Leonard Ayres's (1909) *Laggards in Our Schools* documented high rates of "retardation," by which he meant schoolchildren older than was typical for their grade. Ayres constructed an index of retardation, fancifully based on factory throughput, to measure schools' efficiency at advancing students through grades. Joseph Mayer Rice's 1913 bestseller, *Scientific Management in Education*, reported on the results of the new educational testing and the prospects for making American schooling more efficient.

No corner of American life was safe from the efficiency experts during the heyday of scientific management. In his *Scientific Management and the Churches* (1912), Shailer Matthews (1863–1941), a leading social gospeler and dean of the University of Chicago's Divinity School, proposed to make American churches more efficient, using the methods of Frederick Taylor. Sociologist and minister Samuel Dike (1912) likewise called on the "ecclesiastic engineer" to make church practices less wasteful. If the church wanted society to listen to its message, Dike said, the church must itself hear the message of society.

Efficiency also marched into the American home. Ellen Swallow Richards, the first woman to graduate from the Massachusetts Institute of Technology and an MIT instructor in sanitary chemistry, founded the academic discipline of home economics to promote greater efficiency in the domestic sphere. Richards believed that "the work of home-making in this engineering age must be worked out on engineering principles."[54] Home economics was not just the study of stretching a dollar. Richards conceived of her discipline as a science of "human ecology." She coined the term "euthenics" to describe the science of producing "more efficient human beings" by improving living conditions."[55]

Feminist Charlotte Perkins Gilman observed that housework by unpaid housewives was backward and wasteful, because it was assumed to be women's work without regard to a woman's skill or training, and because, with kitchens and laundries in many homes, it failed to take advantage of economies of scale in domestic production.[56] Cooking, washing, and childcare would be improved and would cost less if produced in central kitchens, laundries, and nurseries. Gilman had many arrows in her rhetorical quiver, but in 1913 she could discredit traditional sex roles by taking aim at their inefficiency.

Human heredity could also be made more efficient. Economist Irving Fisher wrote *National Vitality: Its Waste and Conservation*, one of three volumes produced by the National Conservation Commission, directed by Gifford Pinchot, chief of the US Forest Service and a friend of Fisher's.[57] When Theodore Roosevelt transmitted *National Vitality* to Congress in 1909, he called it "one of the most fundamentally important documents ever laid before the American people."[58]

Fisher's report advocated greater and more efficient federal management of the nation's water, timber, mineral, and land resources. "Conservation," which was a cognate term for efficiency in the Progressive Era, was understood to comprise human resources as well. The Progressive Party's 1912 platform called the conservation of human resources "the supreme duty of the nation."[59]

National Vitality promoted a litany of improvements in public health and hygiene, what Fisher described as conservation in all its branches. But most important of all, Fisher concluded, was conservation of human heredity, by which he meant eugenic regulation to prevent the prolongation of weak lives and to conserve the racial stock. Fisher did not mince words. If it should prove true, he wrote, that "humanitarian impulses betray us into favoring the survival of the unfit and their perpetuation in the next generation, such shortsighted kindness must be checked."[60]

Charles Van Hise, president of the University of Wisconsin and a pillar of the Wisconsin Idea, also made eugenics a keystone of the American conservation movement. In *The Conservation of Natural Resources in the United States* (1910), Van Hise said that Americans must abandon individualism for the good of the race. Individuals were only stewards of their germ plasm, he wrote, holding genetic resources, like land resources, in trust for future generations.

Van Hise demanded that "human defectives" surrender to the state the control of their genetic resources. Whether by involuntary sterilization, or segregation in asylums, hospitals, and institutions, the methods of conserving human heredity, Van Hise warned, must be thoroughgoing. Addressing a visiting delegation of more than one hundred of Philadelphia's leading citizens, who had come to Madison on an "expedition" to study the virtues of the Wisconsin Idea, Van Hise asserted, without qualification, "we know enough about eugenics so that if that knowledge were applied, the defective classes would disappear within a generation."[61]

MEASURING HUMAN RESOURCES

The efficiency experts made a fetish of measurement, which dovetailed neatly with the established fact-finding enthusiasm of economic reformers. The systematic gathering of social and economic facts, quantitative and qualitative, formed the core of the progressives' scientific sensibility. Their empirical investigations, most famously surveys of the harsh conditions of American industrial slums, served multiple purposes: to identify the needs of the urban poor, to create moral outrage, to enlighten and mobilize public opinion regarding the plight of the poor, and also to lend objectivity to reform activism by employing up-to-date scientific methods of investigation. Progress was possible, Simon Nelson Patten said, only with measurable results.[62]

The Hull House settlement workers, led by Florence Kelley and funded by the US Bureau of Labor Statistics, blazed the trail in 1895. Following the example of Charles Booth's London poverty investigations, *Hull House Maps* collected and mapped extensive data on Chicago's immigrant communities. When Kelley and Jane Addams turned to Richard T. Ely for help, he publicized the project and published it as a scholarly work in his influential reform series *The Citizen's Library of Economics and Politics*.[63]

The landmark Pittsburgh Survey of 1907–1908, overseen by Paul Kellogg, editor of *Charities and the Commons*, obtained $47,000 of from the Russell Sage Foundation, roughly $1.1 million in 2014 dollars.[64] The investigation ultimately employed dozens of advisors and field workers, among them

John R. Commons and some of his Wisconsin graduate students, and it produced six volumes.[65] Kellogg soon renamed his journal *The Survey*, a measure of the importance of social investigation to progressive reform.

Arguably the finest social survey was W.E.B. Du Bois's (1868–1963) study of African American urban life in *The Philadelphia Negro* (1899). With minimal funding from the Wharton School, Du Bois did all the work himself.[66] It was a herculean effort with a sophisticated economic analysis. Nowhere in the survey literature was moral outrage more appropriate than in response to the poverty, discrimination, and hatred endured by black Philadelphians, but Du Bois's book was met largely with indifference. White progressives were polite, but they mentally filed Du Bois's scholarship under the "Negro question," treating it as a category separate from their own investigations into the labor question and the immigrant question, even as, ironically, Du Bois made a compelling case for their interrelation.[67]

Government bureaus and temporary commissions entered the social investigation field in fits and starts. An outstanding example, which we examine in Chapter 7, was the Dillingham Commission (1907–1910), chartered by Congress to survey the "immigrant problem." The largest federal investigation yet undertaken outside the US Census, the Dillingham Commission employed a staff of 300 field agents, statisticians, and clerks, and it surveyed more than 3 million immigrants in 300 communities.[68] Its experts produced forty-one volumes, weighing in at nearly 29,000 pages, investigating immigrant households, schools, banks, charity seeking, criminality, head shape, and, first and foremost, the adverse effects of immigrants on American workers' wages and employment.[69]

Government offered efficiency-minded social scientists something that private funders could never match—ready access to human subjects. Captive groups could not say no to the experts who measured their bodies, their character, and their intelligence. School children, the institutionalized, US Army draftees, and immigrants arriving to entry stations were all made available to Progressive Era social scientists, not least the bold pioneers who proposed to measure human intelligence.

Before experts dared to measure intelligence, a complex trait, they measured human heads. At the turn of the century, Thorstein Veblen's *Journal of Political Economy* published an outpouring of articles by economist Carlos Closson, who popularized and proselytized for the scientific racism of two physical anthropologists, Georges Vacher de LaPouge and Otto Ammon.[70]

The anthropologists, who gave their field the grandiose name of "anthroposociology," measured thousands of human heads to calculate a cephalic index, the ratio of head width to head length. In this fashion, they believed they could scientifically demonstrate a permanent race hierarchy. The superior races had longer heads (and thus a lower cephalic index).

The most influential racial taxonomy of the Progressive Era was William Z. Ripley's *The Races of Europe*, published in 1899, a year before President McKinley appointed him to investigate the trusts as a member of the US Industrial Commission. Ripley, an economist trained at MIT and Columbia, spent a long career at Harvard studying transportation economics. In the 1920s he made a name scourging the railroads and Wall Street for dubious accounting and financial practices. His reputation as a financial reformer, cemented during a tour of service with the Interstate Commerce Commission, was such that the *New York Times* published a glowing profile under the headline, "When Ripley Speaks, Wall Street Heeds."[71] Ripley was elected president of the AEA in 1933.

But the young Ripley's first loves were physical anthropology and geography. *The Races of Europe* was a richly detailed, atlas-like anthropological compendium of European ethnicity. It brimmed with maps of Europe detailing the distribution of hair color and head shape, along with mug-shot photographs of ideal racial types. Working together, Ripley and his wife ransacked the physical anthropology literature sufficient to separately publish a 160-page bibliography.[72]

The Races of Europe classified Europeans into three distinct races, using a tripartite scheme: cephalic index, color, and stature. Head, hue, and height distinguished Teutons, Alpines, and Mediterraneans. The northern Teutonic race was long-headed ("dolichocephalic"), tall in stature, and pale in eyes and skin. The southern Mediterranean race was also long-headed but shorter in stature and dark in eyes and skin. The people of the central Alpine race were round-headed ("brachycephalic"), stocky, and intermediate in eye and skin color.

Thorstein Veblen grounded his theory of the rapacious capitalist, made famous in *The Theory of the Leisure Class*, on the varying instincts of Ripley's three European races.[73] Veblen claimed that the capitalist was the product not of his social environment, but of an archaic, predatory race instinct.[74]

By the standards of the day, Ripley and Veblen were, arguably at least, more race scientists than scientific racists. Ripley took account of environmental influences on racial type, as when migration to cities caused changes in stature. Still, he could not disguise his contempt for immigrants without Teutonic blood.

Eastern European Jews, Ripley declared, were the product of a "great Polish swamp of miserable human beings," which, without immigration restriction, would drain itself into the United States.[75] In 1908 he warned of a dual threat to American racial integrity. First was the usual race suicide indictment: inferior immigrants were outbreeding their Anglo-Saxon betters. But Ripley had landed on another hereditary threat: the immigrants were also interbreeding. The mixing of inferior races, he said, threatened to produce an atavistic European type, a kind of negroid throwback.[76] The *New York Times* covered Ripley's alarming forecast under the headline, "Future Americans Will Be Swarthy."[77]

Viewed from the distance of more than a century, Ripley's notions appear both reprehensible and farcical. But Ripley's work was not merely popular, still less was it treated as pseudoscience. On the contrary, anthropologists worldwide regarded *The Races of Europe* as a significant scientific advance. In 1908, the Royal Anthropological Institute of Great Britain and Ireland awarded Ripley the Huxley Medal, its highest scientific honor.

Lapouge and Closson, for their part, boasted that the scientific measurement of heads was destined to "revolutionize the political and social sciences as radically as bacteriology has revolutionized the sciences of medicine."[78] Head shape would provide economics and the other social sciences with a firm scientific foundation. Veblen gave these authors a platform in the *Journal of Political Economy* and appealed to their authority in his landmark criticism of orthodox economics as unscientific.[79]

Later in the Progressive Era, emboldened social scientists began using mental tests rather than head shape to measure human intelligence. The administrative state was vital to their project. It provided human subjects when the experts convinced officials that the identification of mental defectives would significantly reduce the social cost, increasingly borne by government, of "crime, pauperism and industrial inefficiency."[80]

Stanford psychologist Louis Terman created his intelligence test with the idea of detecting the "high-grade defectives" who would prove to be a burden on the state but were passing as normal. Intelligence testing, Terman promised, would expose the tens of thousands of mental defectives passing as normal and bring them under the "surveillance and protection of society."[81]

Psychologist Henry Herbert Goddard (1866–1957), superintendent of the Vineland Training School for Feeble-Minded Girls and Boys in Vineland, New Jersey, was among the first to recognize and avail himself of the administrative state's usefulness to his new discipline.[82] Goddard was well known as the author *The Kallikak Family* (1912), a best seller in the degenerate-family genre. Like Terman, he retailed intelligence testing as an efficient and scientific means of identifying inferiors, and, in 1913, he persuaded immigration officials to let him try out the new techniques at Ellis Island. Goddard's tests were primitive at best and were administered to only 141 subjects—Jews, Hungarians, Italians, and Russians. But his results were startling enough to make headlines; 84 percent of immigrants proved to be feeble-minded.[83]

Goddard's Ellis Island trial was but a warm-up for a vastly larger experiment. Goddard joined forces with psychologist Robert Yerkes (1876–1956), president of the American Psychological Association. Yerkes persuaded the US Army to fund and compel intelligence testing of Army draftees for World War I. Conscripts did not fare much better than had the freshly arrived immigrants.

The Army test results, published under the auspices of the National Academy of Sciences and the US Surgeon General, reported that 54 percent of Army draftees were "morons," that is, high-grade mental defectives with the intelligence of a child aged eight to twelve.[84] Forty-seven percent of white draftees and 89 percent of black draftees were morons, the experts reported, mental inferiors threatening to pass as normal.[85]

The foreign born invariably scored lower than the native born, and immigrants from southern and eastern Europe scored lower than immigrants classified as Teutonic. After the war, when the debate over race-based immigration quotas was especially heated, Princeton psychologist Carl Brigham popularized the Army test results. The data, Brigham informed his readers, clearly demonstrated "the intellectual superiority of our Nordic group over the Mediterranean, Alpine and Negro groups."[86]

With nearly 1.7 million human subjects at their disposal, the experts were free to experiment. At one camp, psychologists graded black soldiers by shade of skin color and found that "true mulattoes" and "lighter negroes" tested far better than did the darker soldiers. More white blood, the experts concluded, yielded higher intelligence.[87] The intrepid psychologists even managed to round up, with help from local authorities, a cohort of prostitutes working nearby. The median prostitute, Yerkes reported, had the mental age of a ten year old.[88]

Ely lauded the Army testing, because it enabled the state to scientifically inventory the fitness of its human stock. We census our farm animals and test our soils, Ely observed. Surely it was no less important to take stock of our human resources, ascertain where defects exist, and apply suitable remedies. Ely allowed that eugenic science was still in its infancy, but, he claimed, "we have got far enough to recognize that there are certain human beings who are absolutely unfit, and should be prevented from a continuation of their kind."[89]

Four years after the war, Ely was still beating the drum. Economic progress, he said, unavoidably left behind large numbers of "absolutely unfit" people incapable of meeting the demands of modern life. The absolutely unfit would plague society until society controlled their breeding. For Ely, the price of progress was eugenics.[90]

Dubious though the tests and testing methods were, the millions of persons subjected to crude intelligence tests demonstrated one result unambiguously. American social scientists had convinced government authorities to fund and compel human subjects for an unprecedented measurement enterprise, carried out to identify and cull inferiors, all in the name of improving the efficiency of the nation's public schools, immigration entry stations, institutions for the handicapped, and military.

PART II

The Progressive Paradox

The reformer is always right about what is wrong.
He is generally wrong about what is right.
—G. K. Chesterton

5

Valuing Labor: What Should Labor Get?

The supreme economic question, John Bates Clark wrote in 1912, was: Is labor getting its due?[1] Politically charged and analytically daunting, the "labor question" encompassed the most compelling economic issues of the Gilded Age and Progressive Era.[2]

The first challenge was to determine what, in fact, labor *was* getting. Reliable data on wages, benefits, and hours were sketchy at best and were often misleading or nonexistent. The second challenge was to understand how wages were determined. Were wages set arbitrarily, by convention or by the whim of the boss? Or did wages bear some connection to market forces of supply and demand?

More daunting still was the third challenge: What *should* labor get? Should workers get whatever the market paid, or were they entitled to a living wage or even to a wage sufficient to support a family? Should a single man be paid less than his identical-twin brother with a family?

Or should workers be valued by what they produced rather than what they consumed? Paying workers the value of their contribution to production, in turn, left dangling the deeply contested matter of what share of output was created by their labor and what share by the other means of production. The question of what labor should get utterly divided scholars, reformers, and politicians, to say nothing of those with a vested interest in the matter—workers and their employers.

Even if the profound disagreement over what labor should get could somehow be resolved, a fourth challenge loomed. If labor was not getting its due—and many believed it was not—what should be done, and by whom? Should the employer be compelled to pay more, so that workers might have a decent wage, or should the burden rightly fall on the taxpayer? A good progressive, Clark said that any solution would have to lie between letting the state do nothing and having it do everything. This was true, but not terribly helpful; the policy space between anarchy and state socialism was enormous.

Should the state empower labor unions to collectively bargain on behalf of their members and simply enforce agreements? Or should the state unilaterally decide what was best for workers and employers and set minimum wages; fix maximum hours; arbitrate wage disputes; regulate safety; and compel insurance against accident, disability, and unemployment? And, if the latter, who was to determine the amount required for a living wage, or a family wage, or a marginal-product wage, and by what methods?

Lurking behind these daunting challenges lay an ancient and still more fundamental question, one which will be at the heart of the remaining pages of *Illiberal Reformers*: was the value of labor, like the value of any other commodity, simply what could be had in exchange for it? Or was the value of labor something determined by factors outside the nexus of market exchange, such as the worker's race, sex, nationality, class, or legal status?

For nearly all of recorded human history, the notion of laborers selling their labor services for wages was nonsensical. Labor was the compelled agricultural toil of social inferiors in the service and under the command of their betters. In the United States, this remained true well into the nineteenth century. The value of labor depended on what the worker was—free or slave, man or woman, native or immigrant, propertied or hireling—not what the worker produced or wished to consume.[3]

Race notoriously demarcated slave and nonslave, but invidious distinctions were made everywhere. Women, children, indentured servants, immigrants, and unpropertied white men were all treated as social or biological inferiors (or both) to propertied Anglo-Saxon men. Who the worker was, as defined by some amalgam of biology, law, and custom, dictated the nature of the work they did and how that work was valued. The value of labor was not determined by what a worker did. What a worker did was determined by his or her social value.

Political historian Rogers Smith called this "ascriptive inequality." If there was a relative political equality among propertied Anglo-Saxon white men, there was a vast political inequality below, based on the inferiority ascribed to blacks, immigrants, women, and the unpropertied.[4] A black man was inferior because he was black. An immigrant was inferior because he was not American. A woman was inferior because she was female. The hired man was inferior because he had no land, tools, or skills.

Smith's topic was political inequality, but the same hierarchy plagued economic life. Like American political life, full participation in American economic life was strictly limited by race, gender, ethnicity, and class.

In this sense, the "labor question" did not begin with the American industrial revolution. The American industrial revolution unsettled an already existing and elaborate labor hierarchy by drawing millions of immigrants, women, and children into the labor force and turning self-employed men into wage earners, changes that shifted the established labor hierarchy's center of gravity, threatening to topple it.

ECONOMICS AND THE ANCIENT ORIGINS OF INFERIOR LABOR

We need to begin at the beginning. For as long as people have recorded their views on economic life, there have been two constants of political economy. The first constant has been to distinguish two opposed methods of economic coordination: market exchange and administrative command. The second constant has been to scorn markets and to esteem administration.

Hostility to trade is as venerable as trade itself. During the roughly two millennia that separate the students of Socrates from Adam Smith's *Wealth of Nations* (1776), the market was scorned as a disreputable demimonde of moneychangers, pawnbrokers, Shylocks, usurers, factors, gougers, hagglers, hawkers, hucksters, jobbers, middlemen, mongers, peddlers, shopkeepers, and scrambling little profiteers. The low regard for the market was conveyed by the epithets given to its participants.

The scholars who maligned markets admired administration. Administration was favored by Greek philosophers grooming tyrants, theologians vindicating the Church, political arithmeticians calculating for lord protectors, the man of the system whispering in the ear of princes and parliaments, and even trading companies, once they had snatched the sovereign powers of the governments that gave them their monopolies.

The term "economics" was coined to name the principles of administering an Athenian agricultural estate, or *oikos*, a self-contained and mostly autarkic socioeconomic unit, ruled by a master, who commanded wife; children; and numerous laborers, slave and free.[5]

The virtuous master, like all male citizen-aristocrats, was expected to administer the city-state, or *polis*, a privilege denied to all others. As his subordinate, the master's wife was relegated to the socially inferior duty of administering the household. She supervised its provisioning and managed the farming, husbandry, and manufacturing carried out by slaves and by intermediate-status workers made up of foreigners and free Athenians without citizenship. Necessary as it was for sustaining Athens, the Greek philosophers regarded administering the oikos as an inferior calling to the art of administering the polis.

If economics were inferior to politics, both forms of administration were deemed superior to the market, the dishonest domain of trade and credit. A few had the temerity to observe that the evils of trade and credit had been useful to the state's business—waging wars and building civic monuments. But the Athenian elite would sooner follow a plow than sully their hands trading goods or lending money.

Labor was a low calling, and trading and moneylending lower still, and yet all three were either essential or very useful to the purposes of state. The polis depended on economic practices its culture disdained. Confronted with this intellectual tension, the Greeks resolved it with a scheme of social, political, and biological hierarchy. The question of labor's value has been entangled with these hierarchies ever since.

The Greeks relegated the dirtiest labor to slaves, who had no choice, indeed *could* have no choice, since nature had fitted them for slavery. Free noncitizens did the rest of the work. Trade and credit were outsourced to other inferiors incapable of virtue, the aliens in their midst. As immigrant races, foreigners and their descendants, lacking the right blood, were permanently barred from politics and land ownership.[6] They were tolerated, however, when the city-state needed their credit.

The Greek model had extraordinary staying power in Europe. Two thousand years later, an aristocracy still monopolized land ownership, ruled the polity and the economy, and claimed supernatural bases for its privilege. Like his ancient counterpart, the modern aristocrat ruled his estate but did not dirty himself with its economic affairs, which were outsourced to social inferiors.

Social class remained identified with productive function. Labor still meant the agricultural toil of slaves, serfs, or tenants in the service and under the command of their betters. Inferiors of middling status—overseers, craftspeople, and merchants—provided management and trade. Jews were tolerated when estates or states needed credit to finance their wars and monuments.

The ancient conception of economics as administration or management also endured. Into the late eighteenth century, *économie politique* referred to the principles of administering a large agricultural estate. France was the king's estate, and it required the management of skilled administrators, men like political economist François Quesnay.

Then everything changed. In 1789, France was no longer the king's estate, and the United States inaugurated its first president. In ejecting and deposing monarchs, the American and French revolutions enacted the radical idea that individuals were invested with divine or natural rights, and that their rulers must answer to them, not the other way around. The philosophers' doctrine that uninvited government command was illegitimate was enlarged by an equally radical claim from political economy, that government command was economically destructive.

The case against administration and for trade was made most famously in Adam Smith's *Wealth of Nations*, published in 1776. Traditional political economy, Smith argued, wrongly regarded the nation-state as a giant self-sufficient oikos, committing the errors of scorning trade as unproductive, thinking that wealth was money rather than the productive potential of people and their tools, and regarding an economy as a national unit to be managed by and for its rulers. These ancient misconceptions throttled the wealth of nations, injuring the poor most of all.

The crucial mistake, Smith argued, was the ancient prejudice that a persons' economic and social value was fixed in an immutable hierarchy at birth. The slave was born to toil, the master to command. The wife and children were born to obey, the husband to command. Women, immigrants, and non-citizens were born to the inferior pursuits of household management, labor, trade, and finance. Only the male aristocrat was born to politics.

Smith deplored human hierarchy as unjust and illiberal. The loftiest philosopher, he wrote, was no better than the commonest street porter, just better trained.[7] Human hierarchy was also economically destructive, precisely because its illiberal prejudices prevented people from specializing in the work they did best.

Smith's system of natural liberty proposed to free every individual "to pursue his own interest his own way, and to bring both his industry and capital into competition with those of any other man, or order of men."[8] When individuals were free to specialize in what they did best, they produced more at lower cost, and the wealth of nations increased. As every prudent family already understood, Smith observed, it was foolish to produce what will cost more to make than to buy.[9] The movement from autarky to specialization was possible, however, only insofar as specializing workers could rely on others to supply the goods they had abandoned producing.

A market made this possible. The expansion of the market permitted greater specialization, which increased labor productivity and generated higher incomes—in a word, progress. A free people free to trade was no evil; it was, rather, the means for reversing two millennia of economic stagnation. A more liberal land was also a wealthier one.

Smith died in 1790, too early in the British Industrial Revolution to see his radical claims borne out. But he lived long enough to see George Washington inaugurated as the first US president, and Smith believed that America might well offer the most fertile soil for his system of natural liberty.

Smith had cause for optimism. The United States in 1790 was a rare thing, a liberal republic. Scarcity, the scourge of classical political economy, seemed absent; land and natural resources were hyperabundant. Its new government promised an improvement over the hopelessly corrupt European regimes. The American founders, fearing government tyranny in all its forms, checked government power by granting the federal government only enumerated powers, dividing its authority among three branches, and reserving significant powers to the states. The US Constitution expressly protected individual rights to life, liberty, and property.

But not for all. The United States was also a slave republic. It offered liberty only to some and citizenship to fewer still. The Civil War and Reconstruction abolished legal slavery, but the old labor hierarchies persisted, in part because the mid-century politics of slavery and emancipation had pushed the liberal and republican labor traditions apart.

Southern apologists for slavery accused the North of "wage slavery." The benevolent plantation master, they asserted, treated his slaves better than the factory owner treated "hirelings." In this telling, the factory foreman was an oppressor worse than the armed and mounted overseer driving field slaves.

Northern abolitionists replied that equating voluntary employment with the dark crimes of chattel slavery was an outrageous false equivalence. The fundamental difference between "wage slavery" and the real thing was self-ownership. When the slaver made another person his property, he violently expropriated all of the slave's liberties. Factory hands were free to leave their employers to work for other firms, to work for themselves, or to not work at all. Free labor can quit. Slaves cannot quit, cannot work for themselves, and cannot choose not to work.

Well after Emancipation and the end of legal slavery, the republican tradition continued to disdain employment as slavish. Radical republicans regarded the wage earner as servile, dependent, and without autonomy. A man dependent on another man for his living was no better than women, children, domestics, sharecroppers, and other inferiors. In the republican view, the wage earner's poverty was a shameful reflection of a defective work ethic, a lack of gumption, drunkenness, or an instinctive servility. There was land free for the taking. Any hardworking, ambitious, sober, and confident white man could quickly improve his station in the land of opportunity.

Radical republicans looked on the liberal freedom to voluntarily enter and exit employment as a pale substitute for their vigorous republican liberty. The wage earner freely entered employment, but once inside, he was still subject to the boss's will. The employee could quit, but only to submit to another boss. Republican liberty demanded complete work autonomy, which, in turn, required ownership not only of oneself but of other means of production—emblematically, the yeoman's small farm or the journeyman's tools. Property provided the economic independence republican liberty demanded, and it also ensured that the republican man—in contrast to the hireling—was fit for citizenship.[10] Republican ideology made the postbellum wage earner as incapable of civic virtue as were the slaves, foreigners, and other landless people of ancient Athens.

And therein lay the rub. American republican liberty was built for a decentralized, sparsely settled agrarian republic with free land (for white men). Republicanism made liberty hostage to every white man obtaining enough land or capital sufficient to become an independent entrepreneur. As the scale of America industrial organization increased, first in transportation, then in production, the entry costs for republican liberty became prohibitively high.

Terence Powderly, head of the Knights of Labor, in 1880 called for nothing less than "to forever banish that curse of modern civilization—wage slavery."[11] Talk of a cooperative commonwealth appealed to the progressive economists in the early days before their professionalization, but Powderly's vision of self-organized worker-owned cooperatives was more a protest than a workable plan of action. Even if workers could collectively organize to raise sufficient capital, were they willing to gamble some of their wages by assuming enterprise risk? And, if the cooperative worker lacked a share of ownership, was he unfit for civic life?

Even more problematic, large industrial organizations required administration. Management by plebiscite was a nonstarter. Workers might elect their bosses, but was it less slavish to choose a boss by ballot than to choose a boss by voting with one's feet? And what if a worker's candidate lost? Did he become slavish the moment the votes were tallied?

These questions proved to be academic. As early as 1870, about 70 percent of Pennsylvania's working population worked for somebody else. In Massachusetts the figure was closer to 80 percent.[12] Americans were sympathetic to republican warnings that the concentration of industrial wealth threatened to corrupt government. And some shared Powderly's view of the capitalist employer as an avaricious "Shylock of Labor."[13] But the other prong of republican ideology, which condemned increasingly large majorities of working Americans as inferiors unfit for civic life, became an insurmountable political liability.

Labor reform had to come to terms with work for wages. But the old habits of thought changed more slowly than did the hurtling economy, and reformers still saw a bit of the slave in the wage earner, no matter how ubiquitous the employee now was. When millions of women and immigrants, long considered to be inferiors, joined the influx into employment, this only reinforced the prejudice that wage earners were not fit to fulfill the duties of citizenship. Having condemned the employee as a wage slave, unfree and unfit for citizenship, whatever sympathy republican ideology had for the working poor was tempered by a lingering disdain for their inferiority.

WHAT SHOULD LABOR GET?

Confronted with the reality of wage labor, the progressive generation of reformers lobbied for a living wage; a shorter day; and healthier, safer work conditions. Workers, labor reformers argued, should get wages high enough to enable them and their families to live decently. Whatever wages the impersonal forces of supply and demand happened to provide, workers were entitled to a standard of living that gave them reasonable comfort, some recreation, and the opportunity to participate in civic life. Moreover, workers' entitlement to a living wage was rightly claimed from their employers.

When Samuel Gompers was asked what labor wanted, he famously responded: "More! More today and more tomorrow; and then ... more and more."[14] How much more was indeterminate, and that was the point. The capitalist employer was grabbing too large a share of the output produced in cooperation with labor.

The Nation's editor, E. L. Godkin, had a very different reply. Godkin opined that the worker was "not entitled to an atom more than the employer is willing to give in a free market."[15] The employer owed his employee payment for services rendered, not a life of reasonable comfort. Godkin said that if the state decided that workers were entitled to a life of reasonable comfort in the form of a living wage, then the burden of paying it should fall on the state, not on the employer who was meeting contractual obligations.

Gompers and Godkin were both opining on what labor *should* get. Gompers said labor's share was unfairly small and should be large, and Godkin said free and nonfraudulent agreements were fair and should be honored. But neither could say what determined actual wages.

American political economy's theories of wage determination were decidedly plural at the turn of the century. The wages-fund doctrine, which imagined a fixed sum of capital to be divided among all workers, had been disposed of, but in its place were different views. One was based on a new economic theory about the relationship between value of labor and what it produced, a late nineteenth-century theoretical innovation led in America by John Bates Clark, later named "marginalism."

Classical economic theory of the early nineteenth century was supply side and laborcentric. To put it crudely, classical political economy said the value of a good was intrinsic to it, embodied in it during the process of its production, and in particular, determined by the labor that went into it. It became known as the labor theory of value. To know the value of a set of horseshoes, for example, one needed to how much of the blacksmith's labor went into it.

Marginalism reversed the direction of cause. Horseshoes were not valuable because they cost labor to produce. What if there were no horses in the land? The blacksmith's labor was valuable only when and insofar as there was demand for horseshoes. Value resided not in the horseshoe itself, but in the eyes of the beholding buyer. The value of the product, then, determined the value of labor that made it.

The marginal productivity theory of wages, developed first in the United States by John Bates Clark in the late 1880s, proposed that workers (in sufficiently competitive markets) were paid wages equal to the value of what the marginal worker contributed to output. The value produced by the last worker hired determined the wage for all workers of an identical grade, and the wage, together with hours worked, determined the worker's income and thus potential standard of living.

As with horseshoes, the value of a person's work was in the eye of the buyer—that is, the employer. The value of labor, Clark argued, was measurable, at least in principle. Measure the value of the additional goods produced by the last person hired, and you knew what a marginal-product or competitive wage was. Accurate or not, Clark's theory was analytically precise. Other progressive wage theories were more vague and thus offered less guidance to wage investigators and regulators.[16] We know, Clark said of his competitive-wage standard, "at what we should aim."[17]

Crucially, Clark also denied that capital was sterile, as Marxian variants of the labor theory of value had it. The blacksmith's tools, anvil, furnace, and fuel added value to his horseshoes. Not all revenue could or should be spent on wages; some had to go to maintain tools and buy fuel and other inputs. Labor was not entitled to all revenue in wages. Here was the limit to Gompers's "more." But anything less than the full value of labor's contribution to output, Clark also said, was exploitation that required state intervention.

Clark's theory got plenty of notice, but its acceptance was slow and piecemeal. A key element of resistance was that many progressives were reluctant to treat wages as a price. If workers were paid wages in exchange for the value of their productive output, then labor was just another commodity whose price was determined by the impersonal market forces of supply and demand. Having conceded that employment was here to stay, many progressives did not wish to concede that labor should be priced, as Beatrice Webb put it, like a mere commodity, "bought by the capitalist in the cheapest market."[18]

These progressives saw a wage not as the price of a contractual exchange, but as a worker-citizen's rightful claim upon his share of the common wealth produced when the laborer cooperated with the capitalist to jointly create it. Labor was not a commodity. The laborer's claim to wages was not a matter of market exchange. The worker's claim, rather, was determined by his political and social standing, an idea, we have seen, with an ancient pedigree.

Clark thus invited trouble when he claimed that marginal-product wages were not only what labor does get (in competitive markets), but also what labor *should* get. As Clark put it, "to every man his product, his whole product, and nothing but his product."[19] On this score, Clark's critics were legion. His favorite former student, Thorstein Veblen, attacked Clark furiously, deriding marginal-product theory as "neoclassical," an epithet meant to imply that it was just laissez-faire dressed up new theoretical costume.[20]

Clark had naively hoped that the public would seize on his conception of marginal-product wages as a natural solution to the question of what labor should get. It did not. And though the profession would later come to see the professional, political, and conceptual advantages in this formulation, his contemporary critics, like J. Lawrence Laughlin, attacked from the right as well from as the left.

So Progressive Era economics remained eclectic on wage determination. Many progressives thought that wages reflected not the value of workers' output, but workers' consumption needs, or standard of living. Workers got paid enough to get by.

All economists, Edward Bemis claimed in 1888, believed that wages have "a strong and almost irresistible tendency to equal the amount necessary to give workmen their usual necessaries, comforts and luxuries." Bemis's former professor, Richard T. Ely, affirmed the same idea in 1894. Wages were determined by "the habitual standard of life" of the wage-earning classes. The notion persisted. No one even pretends, Scott Nearing maintained in 1915, that workers were paid the value of their contribution to the goods they produced.[21]

They had a point. Marginal product was harder to measure than Clark suggested. Did bosses even know what marginal productivity was? Clark's story, moreover, assumed that all workers were of the same grade, that is, were equally productive.

But the living-standard theory of wages had its own theoretical and measurement shortcomings. Did firms really pay their workers a customary rate that reflected the workers' consumption needs rather than their value to the employer? Would not a greedy employer find it profitable to pay a bit more to keep his best, most experienced workers? And if an unmarried man were paid less than his identical-twin brother with a family, why would a greedy employer ever hire the family man?

The measurement problems arose when the state compelled firms to pay their workers a living wage. Which goods should be included among the necessary comforts of life?[22] "More," Gompers would say, but if the state set the legal minimum too high, then many workers would get less, indeed nothing. As we shall see in Chapter 9, the same progressives retailing the living-standard theory of wages also agreed that a minimum wage set too high would idle many workers. In other words, labor productivity *did* matter, even if it was not as determinative as Clark proposed.

In the end, the living-standard theory of wage determination was less a theory than a way of formulating a widespread anxiety, one that lives on, vigorously, into the twenty-first century. The fear was this: if firms can hire whomever they care to, the work will always go the lowest bidder. Insofar as labor productivity was irrelevant, there was a race to the bottom, and the cheapest labor won.

The decent capitalist, the one who wanted his workers to have a living wage, could do nothing to stop it. If the capitalist paid workers a living wage, he could not compete with unscrupulous rivals, who hired low-standard women, children, immigrants, blacks, and the feeble-minded.

The progressives ultimately turned the living-standard theory of wages into a theoretical construct they called "race suicide." Race suicide was an amalgam of late nineteenth- and early twentieth-century anxieties over jobs being outsourced to the lowest bidder and progressive attempts to define an American nationality, both trends intersecting homegrown American discourses on inferiority—racism, nativism, sexism—and all supercharged by the influential new sciences of heredity, Darwinism, eugenics, and race.

To unpack all this, we must first understand the profound influence of Darwinism, eugenics, and race science on American economics and especially on the progressives. This is task of Chapters 6 and 7.

6

Darwinism in Economic Reform

Heredity rules our lives like the supreme, primeval necessity
that stood above the Olympian gods.

Edward A. Ross[1]

It is difficult to overestimate the importance of Darwinian thinking to American economic reform in the Gilded Age and Progressive Era. Evolutionary thought was American economic reform's scientific touchstone and a vital source of ideas and conceptual support. The Wharton School's Simon Nelson Patten, writing in 1894, observed that the century was closing with a bias for biological reasoning and analogy, just as the prior century had closed with a bias for the methods of physics and astronomy. The great scientific victories of the nineteenth century, Patten believed, were "in the field of biology."[2]

SOMETHING IN DARWIN FOR EVERYONE

To understand the influence of evolutionary thought on American economic reform, we must first appreciate that evolutionary thought in the Gilded Age and Progressive Era in no way dictated a conservative, pessimistic, Social Darwinist politics. On the contrary, evolutionary thought was protean, plural, and contested.

It could license, of course, arguments that explained and justified the economic status quo as survival of the fittest, so-called Social Darwinism. But evolutionary thought was no less useful to economic reformers, who found in it justification for optimism rather than pessimism, for intervention rather than fatalism, for vigorous rather than weak government, and for progress rather than drift. Evolution, as Irving Fisher insisted in *National Vitality*, did not teach a "fatalistic creed." Evolution, rather, awakened the world to "the fact of its own improvability."[3]

In the thirty years bracketing 1900, there seems to have been something in Darwin for everyone. Karl Pearson, English eugenicist and founding father of modern statistical theory, found a case for socialism in Darwin, as did the co-discoverer of the theory of evolution by natural selection, Alfred Russel Wallace.[4] Herbert Spencer, in contrast, famously used natural selection, which he called "survival of the fittest," to defend limited government.[5]

Warmongers borrowed the notion of survival of the fittest to justify imperial conquest, as when Josiah Strong asserted that the Anglo-Saxon race was "divinely commissioned" to conquer the backward races abroad.[6] Opponents of war also found sustenance in evolutionary thought. Pyotr Kropotkin argued that the struggle for existence need not involve conflict, much less violence. Cooperation could well be the fittest strategy.[7] David Starr Jordan, president of Stanford from 1891 to 1913 and a leader of the American Peace Movement during World War I, opposed war because it selected for the unfit. The fittest men died in battle, while the weaklings stayed home to reproduce.[8]

Darwin seems to have been pro-natalist, on the grounds that more births increased the variation available for natural selection. Margaret Sanger argued that restricting births was the best way to select the fittest. Darwin's self-appointed "bulldog," T. H. Huxley, thought natural selection justified agnosticism, whereas devout American interpreters, such as botanist Asa Gray, found room in Darwinism for a deity.[9]

It is a tribute to the influence of Darwinism that Darwin inspired exegetes of nearly every ideology: capitalist and socialist, individualist and collectivist, pacifist and militarist, pro-natalist and birth-controlling, as well as agnostic and devout.[10]

Darwinism was itself plural, and Progressive Era evolutionary thought was more plural still. The ideas of other prominent evolutionists (notably,

Herbert Spencer and Alfred Russel Wallace) were also influential in the Progressive Era, both when they accorded with Darwin and when they didn't.

* * * * *

Dorothy Ross has observed that, in the 1880s and 1890s, the triumph of Darwinism transformed the progressive economists' worldview.[11] In broad outline, this is surely correct. Darwinism, however, triumphed only in piecemeal fashion. For example, evolutionary science did not fully embrace natural selection until the 1940s. There was something in Darwin for everyone, but few American social scientists wanted everything in it.

We may describe Darwinism as gradual evolution caused by the natural selection of small, random variations of inheritable traits. Darwin advanced four ideas: evolution, common descent, gradualism, and natural selection.[12]

Take evolution first. Evolution is the idea the world is not constant but rather is steadily changing, so that organisms are transformed in time. All living things, as Darwin concluded in the *Origin of Species*, "have been, or are being, evolved."[13] The concept of evolution was by no means new with Darwin; the term was Spencer's. Nor was the *Origin* the first scientific account to cast doubt on divine creation. But Darwin's version was persuasive, and most American scholars and scientists happily abandoned the notion that all species were immutable, unchanged over time.

Second, common descent is Darwin's theory that every group of organisms is descended from a common ancestor. All animals, plants, and microorganisms ultimately branch back to a single origin of life on earth. The "tree of life," Darwin called it. Like evolution, Darwin's theory of common descent won acceptance among American scholars and scientists earlier than did gradualism and natural selection, which remained minority views during the Progressive Era.[14]

Third, gradualism is the theory that evolutionary change in populations takes place gradually and not by the sudden production of new individual types. Gradualism implies that variations in inherited traits are minute. As Darwin remarks in the *Origin of Species*, nature doesn't make leaps.[15] Organic evolution proceeds very slowly.

Finally, natural selection is a theory of what impels evolution. It says that evolutionary change occurs by means of the production of inheritable variation in every generation. The relatively few individuals who survive to reproduce, owing to their well-adapted combination of inheritable traits, give rise to succeeding generations, and these traits, especially those conducive to increased reproductive success, gradually predominate among members of the species.

Darwin began the *Origin of Species* explaining natural selection by analogy to artificial selection. Nature selects, Darwin suggested, as does a breeder of dogs or pigeons. Darwin came to regret his metaphor of nature selecting, because it wrongly implied variation was purposeful rather than random. At Wallace's urging, he substituted Spencer's "survival of the fittest" for natural selection in the *Origin*'s fifth edition.

If evolution and common descent won wide acceptance in the Gilded Age and Progressive Era, the other Darwinian ideas were vigorously contested, both inside and outside evolutionary science. Particularly controversial were three unsettled questions with important implications for economic reform: Did survival of the fittest drive evolutionary change? Was evolutionary change gradual or rapid? And was evolutionary change random or progressive in some fashion?

When Darwin died in 1882, he did not know how inheritable traits varied or how they were transmitted to descendants.[16] These large explanatory lacunae left ample room for different interpretations. Harvard botanist Asa Gray, for example, was an early champion of Darwin. He arranged for the *Origin*'s first publication in the United States and publicly opposed Darwin's most vocal American critic, naturalist Louis Agassiz.

Gray was also an evangelical Christian, and he filled one gap in Darwin's account with a theistic twist. God was responsible for the beneficial variation of inherited traits, thereby promoting progressive evolution.[17] Gray's genetic variation was purposeful, progressive, and divine in origin.

Here was a fourth unresolved question in Progressive Era evolutionary science: could traits acquired during an organism's lifetime be inherited by its offspring? Darwin said yes. He accepted Lamarck's claim that acquired characters can be inherited, and even offered a theory of it, pangenesis.[18] Alfred Russel Wallace, a hard hereditarian, said no. If a giraffe elongated its neck in an extreme effort to reach edible leaves, it was not the change produced by this effort that was inherited, but rather the genetic predisposition to a long neck that made the effort unnecessary.[19]

Herbert Spencer disagreed with Wallace. A neo-Lamarckian, Spencer led the charge against the hard hereditarians. He took on biologist August Weismann, whose watershed finding in 1889, that mice with their tails cut off do not bear short-tailed progeny, was seen by some as a crucial experiment refuting neo-Lamarckism.[20]

Like Asa Gray, Spencer proposed purposeful rather than random variation, though he made it human rather than divine in origin. Spencer believed that human beings actively adapted themselves to their environments, improving their mental and physical capabilities. Descendants inherited these improved capabilities. Spencer's view was that self-improvement, and therefore race improvement, came from conscious, planned exertion, not from the chance variation and fortuitous adaptation at the heart of Darwinism.[21]

The mechanism of inheritance was intimately connected to another contested question, one with obvious implications for social reform: was evolution progressive? It was progressive in Spencer, with his Lamarckian bootstrapping, whereas, for Darwin (at least some of the time), evolution did not imply progress, only change. Darwin wrote in the *Origin of Species* that he believed "in no law of necessary development."[22] In the *Descent of Man*, Darwin again warned, "progress is no invariable rule."[23]

Elsewhere, however, Darwin contradicted himself. In the penultimate paragraph of the *Origin*, he wrote, "as natural selection works solely by and for the good of each being, all corporeal and mental endowments will tend to progress toward perfection."[24] With evolutionary progress, fitter could mean better. Without progress, fitter meant only better adapted.

Darwin's ambiguity on progress was significant. Conservatives used the Darwin who promised progress to defend the social status quo and to argue that reform might undo social progress. Progressives used the Darwin who promised mere change to reject the status quo and to argue that reform was necessary to ensure that change was progressive, not regressive.

Herbert Spencer was the champion of neo-Lamarckism, and Lester Frank Ward sometimes found himself in the awkward position of defending Spencer, the same man whose individualism and free-market economics he had made his biggest target.[25] It was a price worth paying, for if Darwinism meant drift, the product of chance, then neo-Lamarckism offered progress by means of intelligent direction.

Lamarckian inheritance also let progressives rebut the conservative claim that reform was only a temporary palliative. If improving bad homes improved bad blood, then reform offered the permanent benefit of better heredity. This aspect of neo-Lamarckism, sometimes called "euthenics," was a crucial weapon in the rhetorical arsenal of American reform. Even if pauperism, drunkenness, prostitution, and crime were hereditary diseases, reform did more than treat the symptoms; it also eliminated the disease.[26]

Lamarckian inheritance explains why Irving Fisher and other reformers railed against "race poisons." *Race poisons* were unhealthy behaviors—drinking, smoking, meat eating, promiscuity—thought to injure the "germ plasm" or genetic material.[27] The alcoholic who acquired his disease through weakness and dissolution "poisoned" his genes, transmitting alcoholism to his descendants. If the individual had a right to ruin her own life, that right ended, neo-Lamarckians claimed, when her choices threatened her descendants' heredity.

Lamarckian inheritance enabled Fisher to represent his public health and eugenics crusades as a joint campaign. Tuberculosis, for example, was a public health hazard that also threatened human heredity if one believed, as did the radical economist Scott Nearing, that the children of parents suffering from tuberculosis would inherit weak lungs.[28] As sociologist Charles Richmond Henderson made clear, eugenics was "not limited to selection of parents;" eugenics also included all measures to improve the quality of existing parents.[29]

Progressives departed from Spencer's neo-Lamarckism only on the question of who should improve heredity. Progressives believed that if one individual could improve his or her children's biological inheritance by exercising, eating healthy food, avoiding alcohol and social vices, then socially planned improvements could improve the biological inheritance of an entire generation and all its descendants. Lester Frank Ward, for his part, thought the reform enterprise depended on the inheritance of acquired characteristics and that until such time as evolutionary science conclusively ruled out environmental improvement of human heredity, it was prudent for progressives to "hug the [Lamarckian] delusion."[30] Many did, and evolutionary thought readily accommodated them.[31]

Darwin regarded the slow gradualness of evolutionary change as a fundamental principle of natural history, as did Spencer. This did not suit progressives, of course. But progressives found what they needed in evolutionists who claimed nature did make leaps. T. H. Huxley, for example, saw no reason that inherited variations had to be infinitesimally small, as Darwin supposed. Why couldn't nature produce dramatic mutations and thus accelerate the pace of evolution?

The rate of evolutionary change was of great consequence for any program to improve human heredity. Eugenicists, whether they knew it or not, were logically required to believe that nature could make leaps. Eugenics presumes that desirable traits can be bred into humanity, and undesirable traits bred out, with reasonable dispatch. Eugenics was of little practical use if it took several thousand generations to breed in better traits and breed out worse ones.

Another aspect of evolutionary thought famously fruitful for economists was Darwinian competition. Was competition in nature a model for markets to emulate, or was it a threat to be controlled? Reformers inclined to the latter, sometimes borrowing Alfred Tennyson's phrase—nature red in tooth and claw—to depict American capitalism as brutish or predatory.

Ward represented nature as a threat to be quashed. Humankind progressed, Ward argued, not because of competition but in spite of it. In fact, because it rewarded inferior persons with wealth, a competitive economy prevented "the really fittest from surviving."[32]

Evolutionary science provided a more benign view of Darwinian struggle, however. Wallace, himself a reformer, argued that natural selection maximized "the enjoyment of life with a minimum of suffering and pain." Pyotr Kropotkin's *Mutual Aid* argued that the struggle for existence need not involve conflict, much less violence. Darwin himself insisted that he used the phrase "struggle for existence" in a metaphorical sense meant to accommodate the "dependence of one being upon the other." If the struggle for existence were less a literal struggle than a metaphorical one, economists like Yale president Arthur T. Hadley could claim that economic competition was healthy rivalry, not the glorification of brute force.[33]

Perhaps most central of all was the question of whether natural selection drove evolutionary change. At the turn of the twentieth century, evolutionary science answered "no." Stanford zoologist Vernon Kellogg, one of America's most prominent evolutionary scientists, wrote in 1907, "Darwinian selection theories stand to-day seriously discredited in the biological world."[34] The Progressive Era's relative disregard for natural selection was such that historians of biology refer to the period as "the eclipse of Darwinism."[35]

Eclipsed did not mean dead, and the doctrine of survival of the fittest was deployed to justify the economic status quo. Thomas N. Carver, a conservative Harvard economist, described the laws of natural selection as "God's regular methods of expressing his choice and approval." The naturally selected, Carver declared, were the chosen of God.[36]

In this instance Carver referred to nations, not individuals, which pointed to a further ambiguity in Progressive Era evolutionary thought, whether individuals or species were the unit of selection. Progressives plainly rejected the idea that the captains of industry were fitter individuals, but they found natural selection more congenial when the competitors were nations or races or even the trusts.

Even as he assailed the rich as unfit, Ward defended race conflict as an important cause of progress in social evolution. Ward described race conflict the "sociological homologue of natural selection," and he regarded the struggle of races to be the most important subject in sociology.[37]

Related was the matter of the trusts. Had the industrial behemoths formed in the decade bracketing the turn of the century outcompeted the smaller firms by making better products at lower cost? Or had they achieved their market positions by subverting competition through unfair trade practices, barriers to entry, political favors, or financial market chicanery? The consolidated giants had survived, but had the process that selected them yielded a socially desirable outcome? John Bates Clark, writing with his son John Maurice Clark, made clear the contingent nature of Darwinian fitness:

> In our worship of the survival of the fit under free natural selection we are sometimes in danger of forgetting that the conditions of the struggle fix the kind of fitness that shall come out of it; that survival in the prize ring means fitness for pugilism; not for bricklaying nor philanthropy; that survival in predatory competition is likely to mean something else than fitness for good and efficient production.

Only from a healthy competitive environment, the Clarks argued, can the "right kind of fitness emerge."[38]

Still greater plurality in turn-of-the-century evolutionary thought was introduced by yet another Darwinian idea, Darwin's theory of sexual selection. In *The Descent of Man*, Darwin devised sexual selection to account for apparently maladaptive traits, such as the peacock's outsized tail or the elk's giant antlers, which seemed to hinder rather than promote survival. Darwin argued that though the peacock's unwieldy tail did indeed increase the risk of the peacock not surviving to reproduce, it also, because of its power to attract peahens, conveyed a reproductive advantage.

There were two important consequences for social reform. First, some reformers made use of sexual selection directly. Wallace borrowed it to argue for the hereditary benefits of socialism. Under socialism, Wallace argued, women would be less economically dependent, and thus, like "latter-day Lysistratas," would choose biologically fitter husbands.

In leaving "the improvement of the race to the cultivated minds and pure instincts of the Woman of the Future," Wallace had been inspired by Edward Bellamy's utopian novel, *Looking Backward: 2000–1887*, the 1888 novel that Wallace said converted him to socialism.[39] Wallace was not alone. At the time, Bellamy's book was the best-selling novel in American history. Nationwide, at least 150 Bellamy clubs were founded to discuss and promote its utopian vision. It even inspired a short-lived political party, the Nationalist Party. Thorstein Veblen's wife, Ellen Rolfe, claimed that their reading of *Looking Backward* convinced Veblen to give up philosophy to study economics.[40]

In *Looking Backward*, the woman of 2000 has been liberated by socialism. No longer obliged to find a husband with the means to support her, she selects her mate based on biological fitness. Reserving themselves for only the fittest men, women collectively serve as "judges of the race." Socialism enabled sexual selection, which, Bellamy wrote, selected for the better types, while letting "the inferior types drop out."[41]

Bellamy's conceit, that eliminating private property would purify the race, found a ready audience among socialists and more radical reformers.[42] Sidney Webb, cofounder of the London School of Economics and guiding spirit of Fabian socialism, put Bellamy's idea trenchantly. A free-market economy, Webb declared, leads to "wrong production, both of commodities, and of human beings."[43]

Darwin's sexual selection theory also implied that organic evolution was inefficient, a wasteful arms race that, while perhaps serving the individual, was detrimental to the group. If one could, at a stroke, reduce the size of all peacocks' tails by half, the species survival rate would be increased at no cost to individual reproductive success. The clear implication was that nature was profligate, and humanity could improve on nature's inefficiency by breeding plants and animals.

Progressive Era evolutionary thought was thus plural and unsettled on fundamental questions: whether environment affected hereditary, whether natural selection impelled evolutionary change, whether the individual or the group was the principal unit of selection, whether fitness consisted solely of reproductive success, and whether evolution offered progress or merely change.

This multiplicity permitted social scientists of virtually any outlook to enlist evolutionary ideas in support of their views. And they did. Laissez-faire

Darwinists, for example, could justify economic competition by depicting it as selecting for the fit. William Graham Sumner once called millionaires the product of natural selection.[44] Equally, Reform Darwinists could discredit economic competition by depicting it as selecting for the unfit. "Fitter," Reform Darwinists said, meant only better adapted.

For Reform Darwinists, the captains of industry succeeded only because they had inherited traits, like rapacity and cunning, well adapted to late nineteenth-century capitalism. The struggle for wealth, Edward A. Ross noted drily, "does not bring to the top the intellectual aristocracy."[45]

Charles Horton Cooley, together with Ross perhaps the most influential of the founders of American sociology, concurred. The best type of man, Cooley said, "may be too broadly human for economic success." From a social standpoint, then, it was a good thing that the rich selfishly sacrificed family life for ambition. The relative infecundity of those who pursued wealth, Cooley concluded, was a welcome "elimination of an unsocial type."[46]

Laissez-faire and Reform Darwinists both invoked hereditary fitness to defend completely opposed positions on market competition. There was, in fact, very little disagreement on the virtues of good heredity, just as there was little disagreement on the virtues of economic progress. The disagreement concerned means, not ends: did human heredity, like the industrialized economy, now require state regulation?

Defenders of free markets argued that Spencerian bootstrapping was the best means of improving human heredity. Leave individuals free of government interference, and they will purposefully improve their minds and bodies—natural selection. Socialists like Wallace and Gilman argued that socialism was the road to better heredity. Elevate the economic status of women, and they will do the eugenic work by selecting fitter males—sexual selection.

Progressives said that regulation was the most efficient route to better heredity. Science will determine who is fittest, and state experts will select them by regulating immigration, labor, marriage, and reproduction—artificial selection.

REFORM DARWINISM IN ECONOMIC REFORM

The vexed term "Social Darwinism" was affixed to laissez-faire economics in the middle of the twentieth century, permanently casting Herbert Spencer and William Graham Sumner as the arch-Social Darwinists.[47] There were two problems. Spencer and Sumner were champions of laissez-faire, but neither was particularly Darwinist. Spencer used his own evolutionary ideas, which predated Darwin's, and Darwin only rarely appeared in Sumner's work.[48] The most Darwinian of American social thinkers, Lester Frank Ward, spent years leading the assault on laissez-faire, and he agreed it was "wholly inappropriate to characterize as Social Darwinism the laissez-faire doctrine of political economists."[49]

Hundreds, perhaps thousands of Progressive Era scholars and scientists proudly called themselves eugenicists. As we shall see in Chapter 7, some even wanted to make a religion of it. Not one person, as far as I know, has ever self-identified as a Social Darwinist or claimed to be advocating Social Darwinism. This should have been a clue.

Social Darwinism has always been a term of abuse, an epithet used by critics to discredit ideas they disliked. Critics have disliked different things, laissez-faire in the case of Sumner and Spencer, but "Social Darwinism" was then, and remains to this day, a pejorative.

The second problem was the inference that if laissez-faire were Darwinist, then its enemy, progressive reform, must be not Darwinist. This mistaken inference was bolstered by claims that Darwinism itself was already just Malthusian laissez-faire dressed up in biological costume.[50]

Darwin was indeed influenced by his reading of Robert Malthus, as was the co-discoverer of the theory of evolution by natural selection, Alfred Russel Wallace. Both men cited Malthus's idea of the struggle for existence as a vital source of inspiration. But Wallace was a socialist who argued that capital was "the enemy and tyrant of labour."[51] It seems unlikely that he fashioned survival-of-the-fittest doctrine to justify an economic system he deplored. And critics trying to tar Darwin with a Malthusian brush have miscast Malthus as a partisan for laissez-faire. Malthus, in fact, was a protectionist, a skeptic of industry, and unenthusiastic about immigration.[52]

The progressives were enemies of laissez-faire, not of Darwinism. In fact, they were inveterate Darwinists, deeply influenced by evolutionary thought; they were also drawn in varying degrees to eugenics and race science.

Thorstein Veblen proposed that economics be reconstructed on Darwinist principles. John Dewey claimed Darwin for his version of pragmatism. Richard T. Ely attempted an evolutionary synthesis to explain the joint evolution of society, economy, and humankind. Simon Nelson Patten's *Heredity and Social Progress* (1903), among many other volumes, tried to provide his sui generis political economy with a biological foundation. John R. Commons's anxiety about inferior heredity pervades his *Races and Immigrants*.[53]

Edward A. Ross produced a continual geyser of warnings about the threat of inferior heredity. Charles Richmond Henderson, the University of Chicago minister and sociologist, focused on hereditary degeneracy in his influential social-work study of the defective, dependent, and delinquent.[54] All these projects in progressive social science can be thought of as part of Reform Darwinism.

The Reform Darwinist assault on laissez-faire was first conceived and long commanded by Lester Frank Ward. Ward was glad to fight on the Darwinian front. With training in botany and geology, he believed that Darwin provided more weapons for assaulting laissez-faire than for defending it. When the US government broke the trusts into pieces, Ward taunted critics by observing the trusts had lost a Darwinian contest, so the Social Darwinist must approve of antitrust.

More seriously, it was Ward who masterminded the progressive attack on the doctrine of survival of the fittest with the argument that survival was always relative to environment. That capitalists survived proved only that they had traits well adapted to the Gilded Age, not that those traits were socially desirable. Survival of the fattest should not be confused with survival of the fittest.

A second counterattack was to make society into an organism. US constitutional law had recently decided that limited-liability corporations were legal persons, entitled to some of the same liberties that protected natural persons from the state.[55] Ward's response was to claim the state was an even larger organism, one that encompassed and thus subsumed corporate and natural persons alike.

A third assault was a frontal attack on natural selection itself. Because the survivors of the Gilded Age jungle were unfit, Ward said, society must protect itself against capitalism's dysgenic tendencies. The problem was that natural selection was wasteful, slow, unprogressive, and inhumane. The solution was social selection, which improved upon nature.

SOCIETY AS ORGANISM

When Reform Darwinists wished to argue that society, not just the individual, could be purposeful, they portrayed society as an evolved organism, an idea many of them had first encountered as graduate students in Germany. Economic reformers returned to the social organism metaphor again and again.

Henry Carter Adams said that a society, no less than an individual, has "conscious purposes." Political journalist Herbert Croly described the American society as "an enlarged individual." Some progressives went beyond metaphor. Ely claimed the state was literally an organism. The social organism, Ely insisted, was not a figure of speech but was "strictly and literally true."[56]

Social organism fit nicely with the kind of organic Christianity popular among social gospel progressives. John R. Commons preached that individuals were not separate particles but organs "bound up in the social organism." Christian economic reform was not a matter of saving human atoms, said social gospeler Walter Rauschenbusch, "but of saving the social organism." Jane Addams described her settlement-house work as a humanitarian movement endeavoring to embody itself "in the social organism."[57]

Ely and Ross depicted the social organism in an especially anthropomorphic vein. In Ely's *Introduction to Political Economy*, students learned that "the State is [a] moral person." Social scientists must recognize, Ross wrote in *Social Control*, that society was not just "a bunch of persons!" Society was, rather, "a living thing, actuated, like all the higher creatures, by the instinct for self-preservation."[58]

Woodrow Wilson, when on the presidential campaign trail in 1912, imported the social organism metaphor to justify his argument for a more powerful and more centralized US government, less hobbled by the US Constitution's outmoded eighteenth-century scheme of checks and balances. The trouble with divided government was that "government is not a machine, but a living thing," said Wilson:

> It falls under not the theory of a universe, but under the theory of organic life. It is accountable to Darwin, not to Newton. It is modified by its environment, necessitated by its tasks, shaped to its functions by the sheer pressure of life. No living thing can have its organs offset against each other, as checks, and live.... Living political constitutions must be Darwinian in structure and practice.[59]

Because the social organism evolved, Wilson argued, it should not and probably could not be bound by an unchanging, antiquated set of rules.

The society-as-organism discourse, whether construed metaphorically or literally, complemented progressives' historicist method and reinforced vital progressive intellectual commitments. The progressives' historicist training, which predisposed them to understand economy and polity as products of their histories, dovetailed neatly with their evolutionary thinking.

The organism metaphor captured the progressive idea that a society, unlike a machine, grows and evolves. Looking forward, society's growth and direction can be nurtured, trained, and directed. Looking backward, organisms have a lived past and an evolutionary lineage. To understand the social organism, one had to study its evolutionary history.

The social organism metaphor, especially when anthropomorphized by Ely, Ross, Croly, and others, also lent vital credibility to the reform doctrine of a national administrative state. If society really was a person—possessing a mind, interests, and a conscience—then the problem of determining what 75 million people wanted was vastly simplified.

The social organism metaphor also embedded progressive anti-individualism. A complex organism is *alive*, something greater than the sum of its parts. When actual individuals were deemed organs or cells, they were made subordinate to the social whole. An organism cannot survive internal conflict, as Wilson said.

The label attached the social whole was less important than the idea that it was an evolved, living entity. Whether called state or nation or society, Adams wrote, the thing was an "organic growth and not a mechanical arrangement."[60] Adams thus contrasted the progressives' social organism metaphor to the liberal metaphor of the state as a contractual creation of free individuals who called it into being and could dissolve it as well.

The social organism had a necessary unity, and it was not an inclusive one. The liberal conception of American nationality was inclusive; it admitted any and all who agreed to abide by the social contract's founding principles. But a biological conception of American nationality entailed some kind of evolutionary consanguinity. An organism is constituted by its own cells, and uninvited parasites or microbes were potential threats to its survival.

REFORM AS ARTIFICIAL SELECTION

The progressives' attitude toward natural selection, or survival of the fittest, was a very different matter. As much as they wove evolution and the social organism into their discourse, Reform Darwinists roundly rejected evolution by natural selection. Natural selection, they said, was wasteful, slow, inhumane, and indifferent to progress—moral or other.

According to Ward, the Gilded Age economic order was a jungle. Continuous economic warfare between labor and capital, and between rival capitalists, senselessly wasted resources, showing the folly of thinking free markets could self-regulate. It was absurd to defend market warfare on grounds of survival of the fittest, Ward said, because competition in nature was itself wasteful. After all, Darwinian natural selection worked by random variation and chance adaption, "precisely the reverse of economical."[61]

Wasteful natural selection should not be a model for human society, Ward said. The better model was artificial selection, or breeding. Human control of nature, exemplified by the domestication of plants and animals, was planned, not random, and thus more efficient.[62]

Scientific breeding of plants and animals not only eliminated nature's wastefulness, it improved nature at a faster rate; moreover, it ensured that evolutionary change was not left to blind chance but was made progressive. In other words, artificial selection substituted human mastery for Darwinian drift.

The next, and crucial, step was to assert that humanity's superior management of nature could equally be applied to human society. For Ward, the analogy held good: artificial selection was superior to natural selection in human society just as it had been in nature. He coined the cumbersome term "sociocracy" to name his vision of social improvement by the purposeful administration of society's evolution. "All true social progress," Ward declared, was "artificial."[63]

Ward's sociocracy was the beating heart of Reform Darwinism. The idea that intelligent management of society would improve, direct, and hasten social evolution made it clear that, while progressives rejected survival of the fittest (natural selection), they did not reject selection. On the contrary, progressives departed only on the question of how selection was best accomplished.

Ward intended his argument for the superiority of artificial selection to natural selection to justify social and economic reform. In *The Promise of American Life,* Herbert Croly made his case for national government, arguing that artificial selection (by which he meant expert-guided reform) was superior to natural selection (or laissez-faire). According to Croly, the state had a responsibility to "interfere on behalf of the really fittest."[64] Croly was not being merely metaphorical. His project was to improve human nature by legislation, and the most effectual of all means to improve human nature, Croly concluded, was to improve "the methods whereby men and women are bred."[65]

Ely likewise justified progressive reform on grounds of the "superiority of man's selection to nature's selection." Natural selection gave us weeds as well as nutritious food plants, said Ely, whereas artificial selection, the breeding of plants, improved on nature. The same was true of the human garden, Ely said. Nature, being inefficient, gives us man, whereas society "gives us the ideal man." "The great word is no longer natural selection," said Ely, "but social selection."[66]

Commons also used the terminology, contrasting the artificial selection of economic reform with the natural selection of laissez-faire.[67] America could not rely upon natural selection, Commons warned, because evolution does not always progress.[68] Progress required artificial selection—the administration of society and economy—which Commons claimed was more efficient and humane.

When progressives condemned natural selection as indifferent to progress, they had in mind not only improved efficiency but also moral improvement. Henry Carter Adams's landmark indictment of laissez-faire, *The Relation of the State to Industrial Action* (1887), justified regulation on grounds that economic competition, like competition in nature, was amoral and inhumane. Because markets rewarded the unscrupulous, and good people had to compete with the unscrupulous, markets tempted good people to into bad behavior. The solution, said Adams, was, regulation of industry. Adams' framing of regulation as a defense of Christian morality, protecting the upright from the corrupting effects of market competition, proved irresistible to progressives.

Ely's 1889 textbook quickly followed suit, warning that unregulated markets were forcing "the level of economic life down to the moral standard of the worst men."[69] Commons said that without regulation, all employers were forced "down to the level of the most grasping."[70] Woodrow Wilson also borrowed Adams' framing. Wilson wrote that regulation protected the ethical businessman from having to choose between denying his conscience and retiring from business.[71]

Adams's premise was that, from an ethical standpoint, Darwinian competition adversely selected. The unscrupulous operator hired women and children, because he did not care about the rights of childhood, the claims of family, or the dangers of race deterioration. He thus undercut his rivals who employed men, forcing his rivals to hire women and children or be bankrupted.

Adams's regulatory solution was to bar children and married women from work.[72] Nearly all progressives supported bans on child labor. But the evils of employing married women were less obvious. Was the wage-earning married woman like the child, being exploited by the factory owner? Or was she taking a job that rightly belonged to a man, and or neglecting her obligations to family and race? As we shall see in Chapter 10, progressives made all these claims, often simultaneously.

Not all Americans agreed that Christian ethics required removing women from employment, closing saloons, enforcing blue laws, or banning immigrants based on race. But Adams's framing was a powerful one. If natural selection were amoral and inhumane, then regulation could—in the guise of artificial selection—provide moral uplift as well as greater efficiency.

WHAT THE PROGRESSIVES FOUND IN DARWIN

American economics became an expert policy discipline in the Progressive Era, long before it became a technical discipline. The mathematical and statistical techniques that are characteristic of modern American economics did not acquire meaningful currency until the Second World War. But the progressive economists represented their program as scientific, a claim they founded, in part, on the authority of Darwinism.

Darwinism offered the progressive economists and their reform allies the imprimatur of science, and the example of Darwin as the scientific man of integrity, developing theories through the painstaking accumulation of facts. Moreover, as Donald Bellomy observed, Darwinism seemed to embody the principle of flux and change so characteristic of the turbulent economic times.[73]

Depicting society as an organism was a rhetorical masterstroke. The social organism implied that the American nation was as unified as an organism, with instincts of self-preservation, purposes, and a conscience. The social organism affirmed progressive anti-individualism by subsuming corporate and natural persons alike.

The social organism also lent vital credibility to the progressive idea of the state administering economy and polity for the good of all. What is good for the organism is good for its constituent parts, and is, moreover, easier for experts to ascertain. And an organism has well-defined boundaries. Its native cells belong in it, but foreign parasites and microbes did not belong and were a threat to the organism's health and integrity.

Other Darwinian concepts helped progressives adopt competition while still rejecting Gilded Age capitalism. Darwinian ambiguity on the unit of selection let them criticize competition where they did not like it (in domestic labor and goods markets) and endorse it where they did approve of it (in international relations). A progressive could argue that domestic markets were dysgenic, while Anglo-Saxon manifest destiny was eugenic.

Lamarckian inheritance gave progressives a ready reply to the conservative charge that reform was ineffectual, because it could not affect heredity. With Lamarckian inheritance, improving bad homes improved bad blood.

Darwin's ambiguity on the question of whether evolution resulted in progress or merely change left enough leeway for progressives to claim that society must take charge of its own evolution if it wanted to ensure progress. And even if Darwinian evolution *did* progress, progressives could argue that natural selection was slow, inefficient, and inhumane. The relative disfavor of natural selection among evolutionists allowed progressives to discredit laissez-faire as natural selection, and advocate economic reform as social selection, which was faster, more efficient and more humane.

Similarly, the recognition that the fittest are only those best adapted to the conditions of competition let progressives judge capitalists (and as we shall see, many workers, too) as socially undesirable, precisely because they were so well adapted to the predatory conditions of Gilded Age industrial capitalism. Progressives justified social and economic regulation as the means by which the adverse selection of unrestrained markets could be turned into the beneficial selection of regulated markets.

Darwinism was the master metaphor of late nineteenth- and early twentieth-century American social thought, and it profoundly influenced economic reform. Important as it was, Darwinism should not be conflated with eugenics or with race science, which were historically distinct discourses. As we shall see in Chapter 7, Darwinian ideas were appropriated to justify as well as to discredit eugenics and race science.

7

Eugenics and Race in Economic Reform

Biology now places at the disposal of social workers a mass of knowledge as yet little appreciated which is, however, destined to revolutionize social programs.

Carl Kelsey[1]

SURVIVAL OF THE UNFIT

"Eugenics" derives from the Greek for "well born" and describes the movement to improve human heredity by the social control of human breeding.[2] The concept was ancient. Plato's *Republic* asked why we breed cattle but not humans. The term was minted in 1883 by Francis Galton, a celebrated Victorian Era polymath and cousin of Charles Darwin.[3]

Galton advanced the three governing premises of any eugenic program. First, differences in human intelligence, character, and temperament were due to differences in heredity. Second, human heredity could be improved, and with reasonable dispatch. Human heredity, Galton said, was "almost as plastic as clay, under the control of the breeder's will."[4] And third, the improvement of humankind, like any kind of breeding, could not be left to happenstance. It required scientific investigation and regulation of marriage, reproduction, immigration, and labor.

In other words, eugenics proposed to replace random natural selection with purposeful social selection. As Galton encapsulated it, "what nature does blindly, slowly and ruthlessly, man may do providently, quickly and kindly."[5]

Galton was an elderly man when England began to take him seriously, living just long enough to see a worldwide eugenics movement catch fire. The

American Race Betterment Foundation was established in 1906, one year before the Eugenics Education Society was founded in England. The *American Breeders Magazine*, later renamed the *Journal of Heredity*, began publication in 1910. The state of Indiana passed its forcible sterilization law in 1907, the first of more than thirty American states to do so.[6]

In 1911, Governor Woodrow Wilson signed New Jersey's forcible sterilization legislation, which targeted "the hopelessly defective and criminal classes."[7] Inspired by the slogan "sterilization or racial disaster," Wisconsin passed its forcible sterilization law in 1913, with the support of the University of Wisconsin's most influential scholars, among them President Charles Van Hise and Edward A. Ross.[8] When Charles McCarthy of the Wisconsin Legislative Reference Library queried Ross on the merits of forcible sterilization, Ross pulled no punches: "I am entirely in favor of it," Ross said, and "the objections to it are essentially sentimental and will not bear inspection." Ross assured McCarthy that involuntary sterilization was "not nearly so terrible as hanging a man, and the chances of sterilizing the fit are not nearly so great, as are the chances of hanging the innocent."[9]

In the first three decades of the twentieth century, eugenic ideas were politically influential, culturally fashionable, and scientifically mainstream. The elite sprinkled their conversations with eugenic concerns to signal their *au courant* high-mindedness. As Ross put it, interest in eugenics was almost "a perfect index of one's breadth of outlook and unselfish concern for the future of our race."[10]

The appalling death toll of the First World War quickened eugenic fears. Ross, voicing a sentiment held by many, bemoaned the "immeasurable calamity that has befallen the white race."[11] In 1915, Irving Fisher told the *New York Times* that the European war's greatest cost was not lives lost or wealth destroyed, but the waste of superior heredity. Because Europe was destroying its best genetic material, the duty to protect humankind's future was now thrust upon the United States. More than ever, Fisher declared, America must improve its hereditary resources by banning alcohol, barring immigrants, and segregating or sterilizing the unfit.[12]

Fisher put his money where his mouth was. In the early 1920s he made millions by inventing a Rolodex-type filing device, and he invested the proceeds in stocks. Fisher bought more stock with borrowed money, and a growing fortune freed him to pursue the reform causes closest to his heart. Until the Crash of 1929 ruined him, Fisher estimated that he had personally spent (in 2014 dollars) $10–15 million on crusades for Prohibition, peace, stable money, and improved heredity.[13]

American eugenic thought probably reached its high-water mark during the First World War and the decade after. Between 1914 and 1928, the number of American university courses dedicated to eugenics increased from forty-four to 376, the latter enrolling some 20,000 students.[14] Eugenicist tracts were best sellers. When Samuel Jackson Holmes (1868–1964), a Berkeley zoologist active in eugenics, published his *Bibliography of Eugenics* in 1924, it listed well over 6,000 titles.

Eugenics was a staple of the biology curriculum at all levels. William E. Castle's 1916 *Genetics and Eugenics*, a widely used college text, went into four editions over fifteen years.[15] George Hunter's *Civic Biology* (1914), the popular text used by John Scopes to teach biology to high school students in Dayton, Tennessee, offended creationist Tennessee legislators, because it represented Darwinian evolution as an established scientific fact. Altogether overshadowed during the 1925 Scopes "Monkey" Trial was the textbook's eugenics material, typical of its genre.

Civic Biology warned of the perils of a low and degenerate race, which it described as parasites "spreading disease, immorality, and crime to all parts of this country." Were they lower animals, the text continued, "we would probably kill them off to prevent them from spreading." Society did not yet permit the killing of inferiors, but it could protect itself with other remedies, such as confining them to celibate asylums.[16]

When Scopes was convicted of violating Tennessee's anti-evolution law, Hunter revised his text to expunge the offending references to Darwinian evolution. But Hunter retained and even expanded the eugenics sections. Scopes's defense team sought to enlist America's best-known evolutionary scientists to testify on the side of evolution. Nearly all of these men—David Starr Jordan, president emeritus of Stanford; Princeton's Edwin Conklin; Stanford's Vernon L. Kellogg; and Henry Fairfield Osborn of the American Museum of Natural History among them—were the scientific leaders of the American eugenics movement.[17]

Until the late 1920s, American geneticists supported eugenics or kept their reservations private while welcoming the funding and publicity eugenics generated. Herbert S. Jennings of Johns Hopkins University resigned from the American Eugenics Society in 1925, a year after writing to Irving Fisher that eugenics societies were no place for men of science. In 1927, his colleague Raymond Pearl, once a very active eugenicist, publicly repudiated eugenics in H. L. Mencken's *American Mercury*, an apostasy that received national attention and caused the withdrawal of a job offer from Harvard.

The defections of Jennings and Pearl were significant, but American geneticists distanced themselves from eugenics only gradually. Moreover, the defectors objected not to the idea of social improvement of human heredity, but to the movement's increasingly ugly racism and anti-Semitism.[18]

PREACHING EUGENICS

Modern eugenics was both a scientific and a social movement. The point of investigating human heredity was to understand it sufficiently to improve it. Eugenicists well understood that the social control of heredity—marriage certification, immigration and labor restriction, confinement to asylums and celibate communities, involuntary sterilization—required overhauling popular and scientific attitudes. The good word of eugenics would need to be spread.

Karl Pearson, Galton's biographer and scientific heir, said that Galton envisioned eugenics as a national creed amounting to a religious faith.[19] What, after all, could be more worthy of devotion than the cause that seemed to aim so squarely at human betterment?

Galton was an effective preacher. Nearing the end of his life, he envisioned the moment when public opinion had ripened sufficiently for scientists to declare "a 'Jehad' or Holy War" on all "customs and prejudices that impair the physical and moral qualities of our race." Galton's influential 1904 lecture, given at the London School of Economics to inaugurate the Sociological Society in England, so inspired George Bernard Shaw that the playwright proclaimed, "there is now no reasonable excuse for refusing to face the fact that nothing but a eugenic religion can save our civilization from the fate that has overtaken all previous civilizations."[20]

Preaching the gospel of eugenics came naturally to American eugenicists. Irving Fisher, whose father was a Congregationalist minister, said that to redeem humankind, Americans "must make of eugenics a religion." Fisher told the 1915 Race Betterment Conference, bankrolled by cornflake inventor and eugenicist John Harvey Kellogg, that eugenics was "the foremost plan of human redemption."[21]

Fisher believed that eugenics would reunite science and religion, because eugenics provided a scientific foundation for religious ethics. Eugenics did not simply assert that war was wrong. Eugenics demonstrated *why* war was wrong—war destroyed the best heredity–and thus eugenics was a new and potent ally of morality. Those who objected to eugenics on religious grounds, Fisher said, were like those who had abused Copernicus, Galileo, and Darwin. They would ultimately bend before scientific truth.

Indeed, the dead-enders who said it was wrong to breed men and women like sheep were already a shrinking minority. When we make "eugenics the biggest pillar of the church," Fisher grandly concluded, science and religion, "these two great human interests, will be marching together, hand in hand."[22]

Harvard geneticist Charles Davenport (1866–1944), the acknowledged leader of the American eugenics movement, also expected eugenics to become a kind of religion, and he drew up a creed.[23] It was the text of a speech delivered at the Golden Jubilee of Kellogg's Battle Creek, Michigan, Sanitarium and was titled *Eugenics as a Religion*. In it Davenport offered an eleven-point creed, which included "I believe that I am the trustee of the germ plasm that I carry."[24]

Their proselytizing, organized by the American Eugenics Society, shrewdly exploited the burgeoning mass media to reach ordinary Americans. They installed instructional pavilions at international expositions, attracting millions of visitors, and they staged "fitter family" and "better baby" competitions at state agricultural fairs nationwide. Winning contestants, judged like livestock by local public health officials, were awarded medals inscribed with the motto "yea, I have a goodly heritage," from Psalms 16.[25]

Evangelizers spread the eugenics gospel far beyond the eugenics institutes and laboratories. Eugenic thinking reached deep into American popular culture, traveling through women's magazines, the religious press, movies, and comic strips. The idea of safeguarding American hereditary, with its concomitant fear of degeneracy from within and inundation from abroad, influenced ordinary Americans far removed from the eugenics movement's professionals and publicists.

A doctrine of human hierarchy, eugenics caught on even faster among the elite. In *The Great Gatsby*, F. Scott Fitzgerald has Tom Buchanan, the personification of established wealth, tell Nick Carraway that civilization is going to pieces, and "if we don't look out the white race will be—will be utterly submerged. It's all scientific stuff; it's been proved."[26] Buchanan's authority was "the *Rise of the Colored Empires* by this man Goddard," a likely allusion to Lothrop Stoddard, whose 1920 best seller was titled *The Rising Tide of Color against White World-Supremacy*.

Jack London plumbed the sensibilities of a very different class, but his work was rife with eugenicist and white supremacist notions. Eugene O'Neill brought hereditarian themes to the American stage, leaving a permanent stamp on American theater.[27] Virginia Woolf confided to her diary that imbeciles "should certainly be killed." T. S. Eliot studied eugenics closely and agreed that segregating and sterilizing "defectives" was necessary to protect society. The human race, Eliot wrote, can, if it will, improve indefinitely, "by social and economic reorganization, by eugenics, and by any other external means possible to the science of intellect."

In 1908, D. H. Lawrence, with horrible prescience, indulged in an extermination fantasy:

> If I had my way, I would build a lethal chamber as big as the Crystal Palace, with a military band playing softly, and a Cinematograph working brightly, and then I'd go out in back streets and main streets and bring them all in, all the sick, the halt, and the maimed; I would lead them gently, and they would smile at me.[28]

Lawrence did not invent the idea of euthanizing inferiors in lethal chambers. It already had some limited currency among American eugenicists.[29]

Before the Second World War, eugenics ideas profoundly influenced Britain, the United States, and Germany.[30] They also thrived in nearly all non-Catholic Western countries and in many others.[31] As historian Frank Dikötter observed, eugenic ideas were integral to "the political vocabulary of virtually every significant modernizing force between the two world wars."[32]

As with Darwinism, eugenics was a very broad church. At bottom, eugenics was based on the fear of inferiority, of being inundated from without, or of suffering degeneration from within. The eugenic threat thus changed with the times, and it also varied from place to place, as did the theories of how inferiority worked to undermine heredity.

Latin American countries tended to be neo-Lamarckian with respect to heredity.[33] Immigrant-receiving countries made nationality central, while British eugenicists focused their fears on the working classes, what Sidney and Beatrice Webb called the industrial residuum. Sweden's eugenicists purported to be nonracist, concerned only with cost burden of mental defectives to its new welfare state, while race was integral to eugenic thought in many places.

Eugenic programs gave scientific authority to a wide variety of experts, depending on where the threat was thought to reside. In some places, historian Mark Adams observed, eugenics "was dominated by experimental biologists, in others by animal breeders, physicians, pediatricians, psychiatrists, anthropologists, demographers or public health officials."[34]

The appeal of eugenics crossed ideological boundaries as well. Early twentieth-century eugenics, like Darwinism, found exponents of every political stripe—conservative, progressive, and socialist alike. In the infamous *Buck v. Bell* (1927) US Supreme Court decision, the conservative William Howard Taft and the progressive Louise Brandeis joined Oliver Wendell Holmes's majority opinion, which announced that the same state power that justified compulsory vaccination was "broad enough to cover cutting the Fallopian tubes." Taft and Brandeis endorsed Holmes's notorious declaration, "three generations of imbeciles is enough."[35] Holmes later confided to his longtime correspondent Harold Laski, a Fabian socialist also taken with eugenics, that when he upheld the law requiring the sterilization of imbeciles like Carrie Buck, he felt he "was getting nearer to the first principle of real reform."[36]

Prominent American eugenicists, including movement leaders Charles Davenport and Madison Grant, were conservatives. They identified fitness with social and economic position, and they also were hard hereditarians, dubious of the Lamarckian inheritance clung to by progressives. But as eugenicists, these conservatives were not classical liberals.

Like all eugenicists, they were illiberal. Conservatives do not object to state coercion so long as it is used for what they regard as the right purposes, and these men were happy to trample on individual rights to obtain the greater good of improved hereditary health.[37]

The conservative eugenicists were not Social Darwinists defending the status quo. The raison d'être of eugenics was to reform the status quo. Conservative eugenicists, no less enthusiastically than socialists and progressives, upheld the use of state power to restrict individuals' rights to marry, reproduce, immigrate, and work.

Historians invariably style Madison Grant a conservative, because he was a blueblood clubman from a patrician family, and his best-known work, *The Passing of the Great Race*, is a museum piece of scientific racism. But Grant's eugenic ideas originated from a corner of the conservative im
connected to Progressivism: conservation.

Grant was a cofounder of the American environmental movement, a crusading conservationist who preserved the California redwoods; saved the American bison from extinction; fought for stricter gun control laws; helped create Glacier and Denali national parks; and worked to preserve whales, bald eagles, and pronghorn antelopes. Grant also opposed war, had doubts about imperialism, and strongly supported birth control.[38]

Like Theodore Roosevelt and Charles Van Hise, president of the University of Wisconsin, Grant regarded eugenics as conservation of the race, thus as part of the same project as conservation of endangered lands and species. In Grant's view, the white Nordic race was endangered, and it was the task of conservationists to ensure that it was preserved. Because individuals could not be trusted with their genetic resources any more than they could be trusted with land resources, the state must take control of both.

Eugenics also counted many supporters on the left, from Fabian socialists like George Bernard Shaw and Sidney and Beatrice Webb to birth control advocate Margaret Sanger, who convinced skeptical eugenicists that birth control could be a valuable tool of eugenics.[39] This was no small feat of persuasion. Many eugenicists feared unregulated birth control was dysgenic in its effects, because, as progressive sociologist Charles Horton Cooley warned, the "intelligent classes" used it, and the inferior classes did not. If the state delivered birth control to the inferior classes, Cooley noted, then contraception could indeed work eugenically.[40]

Radical economist Scott Nearing, author of several eugenicist tracts, asserted that "persons with transmissible defects have no right to parenthood and a sane society in its effort to maintain its race standards would absolutely forbid hereditary defectives to procreate their kind."[41] Biologist Hermann Mueller, who won a Nobel Prize for mutating genes with radiation, argued that, because capitalism rewards the unfit with wealth, a socialist revolution was necessary to be able to scientifically distinguish the fit from the unfit.[42]

With utterly different conceptions of the social good, conservative and socialist eugenicists bemoaned survival of the unfit. Progressive Era eugenics required agreement upon three things only—the primacy of heredity, human hierarchy rather than human equality, and the necessarily illiberal idea that human heredity must be socially controlled rather than left to individual choice.

American eugenics lost some of its luster in the1930s, partly because of its own handiwork, the race-based immigration quotas of the 1920s, which all but ended immigration from eastern and southern Europe and from Asia. When the national birth rate fell below replacement level during the Great Depression, from 21.3 live births per 1,000 persons in 1930 to 18.4 per 1,000 persons in 1933, a shrinking population made differential fertility less threatening.[43]

Still, eugenic policies moved forward with powerful inertial force nonetheless. Over 20,000 Americans were forcibly sterilized by between 1931 and 1939, more than triple the number sterilized between 1920 and 1929.[44]

EUGENICS IN ECONOMIC REFORM

Economic progressives, as discussed in Chapter 6, used artificial selection or social selection to describe economic reform and its superiority to laissez-faire as natural selection. Just as the plant or animal breeder outdid nature, the analogy went, so too did the intelligent administrator of the economy outdo the economy left alone. But "artificial selection" was not just a metaphor for the superiority of administration to laissez-faire; it also referred, more narrowly, to eugenics, the social administration of human heredity.[45]

Many progressive economists followed Irving Fisher in believing that reforms regulating race poisons, such as alcohol, tuberculosis, and sexually transmitted diseases, would improve heredity as well as public health. John R. Commons was this sort of neo-Lamarckian. Heredity, he said plainly, "can be modified by modifying environment." The unhealthy environment of American industrial cities, Commons claimed, was breeding a new race of humans, plagued with "inherited traits of physical and moral degeneracy suited to the new environment of the tenement house, the saloon and the jail."[46]

Like Fisher, Commons used neo-Lamarckian logic to justify reform, while also advocating more direct measures. Uplift was socially costly. Eugenics was cheap. When society prevented the unfit from being born, there were fewer of them to uplift. An enlightened society, Commons said, will "*displace* the baser elements," just as the Anglo-Saxon had displaced the American Indian.[47]

In enumerating hereditary dangers, Commons carefully distinguished race degeneration (what happened to a given race under adverse conditions) from race suicide (what happened when the superior—by which Commons meant Anglo-Saxon—race was outbred by its more prolific inferiors,

African Americans, Asians, and southern and eastern Europeans). Commons also distinguished between race inferiority, which he said was hereditary, and backwardness, which was environmental.[48]

In Commons's view, backward Appalachian whites, owing to their racial fitness as Anglo-Saxons, could be educated and thereby assimilated into American life. African Americans, however, could not be uplifted. Black hereditary inferiority, Commons asserted, could be remedied only by interbreeding with superior races, what he called "amalgamation."[49]

In addition to the 12 percent of Americans who were African American, Commons estimated that nearly 2 percent of the US population was irredeemable defective, mentally or physically.[50] So by Commons's reckoning, 14 percent of Americans—about 10 million people in 1900—were condemned by their heredity to permanent inferiority.

Simon Nelson Patten was the preeminent figure at the University of Pennsylvania's Wharton School in the Progressive Era, a founder of the American Economic Association (AEA) and of the American Academy of Political and Social Science, whose journal, *The Annals*, he edited for years. His influence in progressive circles derived in part from the many talented students his reform activism attracted to Philadelphia, such as Scott Nearing, who later joined the Wharton faculty; Henry Rodgers Seager, Columbia University professor and pioneering advocate of social insurance, and Edward Devine, a cofounder of the social-work profession. Rex Tugwell and Frances Perkins, reformers and leading New Dealers, were students of Patten's, as was Walter Weyl, founding editor of *The New Republic*. In 1911, Patten described himself as someone who has "been fighting for twenty years as an insurgent in economics."[51]

Patten was quintessentially progressive in his social gospel impulses, his biologically informed social science, and his ambivalence toward the poor. Patten saw in American industrialization the beginning of a new civilization. But abundance, by easing the struggle for existence, enabled "the continuance of the low social classes," and the "survival of the ignorant."[52]

Patten put somewhat less emphasis on inherited debility than did Commons or Ross. But unsure whether race "degeneration was due to bad environment or to heredity," he vacillated.[53]

In 1911, Patten decided that cultural transmissions mattered more than heredity for social progress. That same year, however, he equivocated, asserting that heredity was the only thing that could "transform man into a superman and we must rely on it to reach this higher level." In 1912, Patten offered a model of interactions between "round faced" and "long faced" human types, using the cephalic-measurement language familiar to his readers and noting the challenge for the eugenicist of deciding which traits should be bred out. But in 1915 he again contradicted himself, declaring, "eugenics is giving us a stronger man and a vigorous woman."[54]

Patten's ongoing ambivalence about eugenics is perhaps a measure of the quintessential tension in progressive thinking, which simultaneously regarded the poor as victims deserving uplift and as threats requiring restraint.

RACE SCIENCE

Darwin regarded fitness as the outcome of a selective process. Darwinian fitness is determined only retrospectively. Eugenics, however, is premised on the survival of the unfit, so eugenics requires that the fittest be determined before the selective process. Because eugenics *begins* with a hierarchy, it also must postulate who decides what the hierarchy shall be, that is, who determines the fitness ranking that will guide society's selection of the fittest.

In the Progressive Era, it was understood that human hierarchy was a matter for science to determine. Scientific experts ranked groups from best to worst. Race scientists invariably located African Americans at the bottom of their pyramids of humanity. American racism, as Robert Wiebe put it, "worked best in color."[55]

In defining race, American race science was as protean as was evolutionary thought. Eugenicists and race theorists used "race" to refer to the human race as well as to the conventional division of humanity into "white, black, yellow, brown, and red races."[56] They also used the term to describe ethnicity or nationality, especially when cataloguing the different races of Europe coming to America, as we shall see in Chapters 8 and 9.

What was more, "race" overlapped ambiguously with cognate terms, such as "type," "stock," "group," and "people."[57] When race scientists or eugenicists bemoaned the survival of unfit races, they could be describing African Americans, French Canadians, the Irish, Italians of the Mezzogiorno, Russian Jews, the Chinese, or any number of peoples deemed a threat to Anglo-Saxon race integrity. Anglo-Saxonist scholars spent countless hours disputing the question of whether Norman blood had degraded the Anglo-Saxon race.[58]

In whatever sense Progressive Era eugenicists used "race," whether in the context of the "Negro question" or the "immigration question," the survival of the unfit races did not exhaust their concerns. Their catalog of inferiority was far larger. Progressive Era eugenics also promulgated hierarchies of gender, class, intellect, and moral character.

Progressive economists focused their ire on the competition of the immigrant races in the labor market. Until African Americans migrated in large numbers to industrial jobs in northern cities, they received less attention as wage threats. Outside the South, economic reformers typically treated the Negro question as distinct from the labor question.

When progressive economists did consider African Americans, scientific racism was the norm.[59] The AEA published Frederick Hoffman's *Race Traits of the American Negro* in 1896, the same year the US Supreme Court's infamous *Plessy v. Ferguson* decision promulgated its "separate but equal" doctrine, upholding the constitutionality of Jim Crow legislation. Hoffman was an actuary for Prudential Life Insurance and a member of the American Association for Labor Legislation's (AALL) Administrative Council.

Hoffman said that American blacks, freed from slavery, had become "lazy, thriftless, and unreliable" and were soon to "attain a condition of total depravity and utter worthlessness." African Americans, Hoffman concluded, were doomed to extinction, victims not of Jim Crow's appalling conditions but of black hereditary inferiority, especially a low standard of sexual morality, a "race trait of which scrofula, consumption and syphilis are the inevitable consequences."[60]

Even as Hoffmann claimed that black hereditary inferiority condemned African Americans to extinction, he warned that, until that time, African American were a serious hindrance to white economic progress. Education, philanthropy, and religion had all failed to inculcate in black people the "stern and uncompromising virtues of the Aryan race." The white race, Hoffmann warned ominously, "will not hesitate to make war upon those races who prove themselves useless factors in the progress of mankind"[61] Richmond Mayo-Smith, the Columbia University economist, took a less militant view than did Hoffman, but he too condemned black Americans as lacking the intelligence and virtue necessary for full equality in American political and social life. "The negro" will "always be a problem for us," Mayo-Smith asserted, but their "docility and good nature" made African Americans "a comparatively harmless, if not a progressive and desirable, element in our national life." Ely was no less condescending. "Negroes," he wrote, "are for the most part grownup children, and should be treated as such."[62]

Commons, for his part, described African Americans as "indolent and fickle," notwithstanding the fact that blacks did the work deemed too difficult or dirty for white workers.[63] Black laziness explained why, Commons argued, their enslavement was defensible, even necessary: "The negro could not possibly have found a place in American industry had he come as a free man.... [I]f such races are to adopt that industrious life which is second nature to races of the temperate zones, it is only through some form of compulsion."[64]

Ross was no better disposed to African Americans than were his Wisconsin colleagues. More than forty years after the end of the Civil War, Ross wrote: "The theory that races are virtually equal in capacity leads to such monumental follies as lining the valleys of the South with the bones of half a million picked whites in order to improve the conditions of four million unpicked blacks." The black "millions of inferior race," Ross said, dragged down American energy and character.[65]

Charles Cooley, the eminent sociologist, warned that providing better healthcare and nutrition for African Americans would lower black death rates, raising the dysgenic specter of the black population "overwhelming" the white. Fearing a rising tide of colored people, Cooley proposed that improvements in African American healthcare be accompanied by eugenic measures designed to reduce the quantity and improve the quality of black births.[66]

Among the progressives, John Bates Clark makes for a refreshing and illuminating contrast. Like nearly all progressives, Clark made use of evolutionary ideas. But he shunned the hereditarian thinking of Ely, Commons, Cooley, Fisher, Patten, Ross, and the others caught up in race science and eugenics.

Clark made no attempt to judge markets by their putative consequences for human heredity. When he spoke of the salutary effects of survival of the fittest, he was referring to competition among business firms, not races.[67] Clark argued for the racial equality of African Americans, insisting that black success depended not on heredity but on the opportunity to own land.

At the 1891 Mohonk conference on the Negro Question, Clark observed, "a part of the difficulty lies, probably, in the Negro's psychology; but that is not so deeply rooted that it cannot be eradicated. It is not, at any rate, permanently in the blood."[68]

Clark envisioned African Americans acquiring their own farms with the end of unjust restrictions, and, as fully propertied farmers, competing successfully on an equal economic footing with Southern whites.[69] At a time when most race scholars forecast black decline and even extinction, Clark predicted and welcomed African American economic success.

African American scholars, such as Kelly Miller (1863–1939) of Howard University, rejected the racism of their progressive white peers. But some also warned of dysgenic trends in the black community. Miller was the first African American to attend Johns Hopkins (1887–1889), and he was a contemporary of the legion of progressive social scientists and activists taught by Richard T. Ely. Miller studied mathematics, physics, and astronomy, his mathematical gifts having been recognized by Simon Newcomb, who commended him to the University.

Miller worried that ordinary blacks were outbreeding "the higher element of the Negro race," by which he meant the Howard faculty. The average Howard professor, Kelly wrote, produced less than one child, potentially threatening the extinction of the "Talented Tenth," the term used by W.E.B. Du Bois to describe the educated elite among African Americans.[70]

Du Bois was a tireless enemy of American racism, not least when race scientists identified race with inferior heredity. He discredited purported racial differences in intelligence, rigorously dismantling the studies that used cranial measurements to allege black intellectual inferiority. In a 1904 lecture on heredity, Du Bois assailed the notion that race was connected in any way to hereditary fitness. There was no proof, and probably no possible proof, Du Bois said, for the thesis that "the physical development of men shows any color line below which is a black pelt and above which is a white."[71]

Du Bois debunked racial inferiority, but he accepted—indeed, warned of—hereditary differences across individuals. There were degenerate types among blacks, Du Bois said, just as there were among Europeans. And as much as black Americans suffered disproportionately and unjustly from adverse social conditions, heredity mattered too. Of a million black youth, he wrote, "some were fitted to know and some to dig."[72]

Du Bois's parsing of human difference reveals two subtleties of American eugenic thought in the Progressive Era. First, racism, its ubiquity and special salience notwithstanding, was neither necessary nor sufficient for eugenic thought. American eugenics was deeply racist, but it took other forms as well. Second, Du Bois helps illustrate the extraordinary sway of hereditarian and hierarchical thinking on American scholars during the Progressive Era.

One of America's leading opponents of scientific racism felt compelled to warn that the unfit were outbreeding their betters.

The "degenerate families" literature begun by Richard Dugdale's *The Jukes* also located hereditary inferiority in class rather than in race. The "white trash" families given pseudonyms such as the Jukes, the Kallikaks, the Nam, or the Tribe of Ishmael, were mostly of Anglo-Saxon background.[73] They were judged deficient in intellect and morals, which explained their feeble-mindedness, promiscuity, dissolution, and pauperism. As Irving Fisher put it, the threat to Anglo-Saxon race integrity "need not come from the outside, it may come from inside the decadent nation."[74]

It was a commonplace among eugenicists to portray the poor—Anglo-Saxon or otherwise—as paupers. Anxious that public assistance promoted survival of the unfit, Edward A. Ross referred to poor relief as "maleficent charity."[75] The cult of charity, Ross warned, had "formed a shelter under which idiots and *cretins* have crept and bred."[76]

With fitness conceptually untethered from survival, it was also possible to bemoan the "idiots and cretins" among the rich. Thorstein Veblen mercilessly lampooned the conspicuous consumption of America's Gilded Age leisure class. Veblen's view was that capitalists produced nothing of value and then spent fortunes on equally worthless goods to parade their wealth. How had such useless fools prospered? Veblen found his answer in heredity: the capitalist was able to exploit everyone else because he had inherited an atavistic, predatory race instinct.[77]

Ross, casting his jaundiced eye upward, also found deficiency among the decadent rich. Shielded by wealth from the rigors of natural selection, the fools and weaklings among the rich, Ross warned, were free to "propagate their kind unhindered."[78] Ross likely picked up the idea from Darwin, Wallace, and Francis Galton, all of whom accused English primogeniture laws of producing feeble-minded older sons among the aristocracy.[79]

David Starr Jordan, the Stanford University president, concluded that England's loss was America's gain. Denied land, the younger sons of the English nobility, Jordan wrote, became the Roundhead, the Puritan, and the Pilgrim on the Mayflower. Every Anglo-American thus had his veins warmed with noble and royal blood. Genealogical studies, President Jordan wrote, demonstrated that all this was "literally true."[80]

Of course, the threat of the unfit rich was invoked less often than the threat of the unfit poor. The poor were far more numerous, allegedly produced larger families, and were more likely to be of a race already deemed inferior. But the survival-of-the-unfit-rich argument reminds us of how influential the discourses of human biological inferiority were during the Progressive Era, and of how readily hereditarian claims were grafted on existing modes of argument.

A left eugenicist could discredit capitalism by invoking the specter of hereditary feebleness among the decadent rich, just as a right eugenicist could defend capitalism by making wealth the just deserts of superior fitness. The enemy of American racism could warn of the dangers of heredity inferiority.

Racism, bigotry, nativism, and suspicion of wealth and position were nothing new to America in the Gilded Age and Progressive Era. All were hardy perennials of the first century of the republic. But hereditarian thinking was new, or at least newly scientific in its presentation.

The new discourses of eugenics and race science recast spiritual or moral failure as biological inferiority, making old prejudices newly respectable and lending scientific luster to the arguments of critics and defenders of American economic life.

THE ANGLO-SAXON RACE

Eugenics wanted more from the fit and less from the unfit. In the United States, the unfit got far more attention, with eugenicists creating a vast and varied catalog of inferiors, immigrants, blacks, women, the disabled, and other types. Like racism, American eugenics largely focused its fears on the unfit, giving less direct attention to the Protestant Anglo-Saxon men the unfit were allegedly threatening.

The privileged position of this group was largely unexamined, just as "whiteness" was largely defined by what it was not. Still, some progressives scrutinized their own group. Theodore Roosevelt demanded more children from the fitter classes and races. And other progressives valorized their Anglo-Saxon heritage as the hereditary bearer of American virtues they purported to preserve.

Walter Rauschenbusch, the radical social gospeler, found his cooperative commonwealth in the Aryans of the ancient Saxon forests. Francis Willard, of the Women's Christian Temperance Union, located temperance in Saxon heredity, and drunkenness elsewhere. The first members of the history profession in America, led by Herbert Baxter Adams (Ely's senior partner at

Johns Hopkins), made the capacity for liberty and self-government Anglo-Saxon race traits.

Rauschenbusch, the most influential social gospel theologian of the twentieth century, described the social gospel as a translation of evolutionary theory into religious faith, and he placed heredity at the center of economic reform.[81] To "Christianize" society, Rauschenbusch preached, the state must eliminate free markets, a "murderous" system ruled by the "law of tooth and nail," and replace it with a cooperative commonwealth of comradeship and solidarity.[82]

Rauschenbusch's cooperative commonwealth was a "fraternal democracy" that shared all property in common. The capacity for fraternal democracy he made hereditary. Fraternal democracy, Rauschenbusch claimed, had its evolutionary origins in the early history of "our Aryan race." Cooperation and common property were "dyed into the fiber of our breed," innate to the Anglo-Saxon.[83]

The evils of industrial capitalism thus were not native to Anglo-Saxon America, but were imported by immigrants from the south and east of Europe, who, by undercutting American wages, shrank the "Teutonic stock," and with it, the American capacity for fraternal democracy and common property. Capitalism drew its ever-increasing strength from the survival of the unfit immigrant.[84]

Rauschenbusch's solution to murderous capitalism was to eliminate its sustenance, the unfit. We know enough, he claimed, to direct human evolution. Let us "make history make us."[85]

Rauschenbusch spent eleven years ministering to an immigrant congregation in New York City's Hell's Kitchen neighborhood. He must have felt some compassion for his destitute flock. But he also regarded them as a menace, ultimately offering them the open hand of spiritual uplift and the closed fist of racial exclusion.

Rauschenbusch's mishmash of the social gospel, economic reform, Anglo-Saxonism, Darwinism, and anti-Catholicism was motley but not unprecedented. A well-known version was already available in Josiah Strong's best seller, *Our Country*. Anglo-Saxonism, like Darwinism, was adaptable, and other scholars and activists adopted it when it suited their purposes.

Frances Willard appealed to Anglo-Saxonism as part of her crusade against alcohol. Leader of the influential Women's Christian Temperance Union (WCTU), which by the early1890s boasted 150,000 members, she was courted by other reformers and made alliances with labor and woman's suffrage groups. Willard ranked her reform commitments when she announced, "I am first a Christian, then I am a Saxon, then I am an American."[86] It was an echo of the WCTU's motto: "For God, Home and Native Land." In the 1890s, she found a common enemy for her commitments, Catholic immigrants. Catholics were neither Protestant, nor Saxon, nor American by birth, and most rejected as bizarre the notion that drinking beer and wine should be criminal. Worse yet, Willard charged, Catholic immigrants undercut American workers and clung to retrograde views of women.

Willard's reform remedy was race-based immigration restriction and alcohol prohibition. Stringent immigration laws, she said, would stop the influx of "the scum of the Old World." Ridding America of alcohol and the wrong kind of immigrants would restore its hereditary greatness. We can, Willard told her followers, "weld the Anglo-Saxons of the New World into one royal family."[87]

Anglo-Saxonism, which was politically malleable, attracted conservatives as well. Columbia University's John Burgess was an influential Teutonist, as was Senator Henry Cabot Lodge. Lodge, the first student to obtain a PhD in history from Harvard, wrote his doctoral dissertation on the German origins of Anglo-Saxon land laws and later joined the anti-immigrant cause in the name of preserving Anglo-Saxon race integrity.[88] Another Anglo-Saxon enthusiast was John Fiske, who popularized Herbert Spencer's evolutionary ideas and who believed that Anglo-Saxons, uniquely, could merge with other peoples and still preserve their superior race traits.

Conservatives like Lodge seemed a world apart from social gospelers in the mold of Strong, Rauschenbusch, and Willard. Lodge opposed woman's suffrage and regarded Prohibition as dangerous. But, whether patrician or evangelical, they all embraced the notion that what was good in America was the product of its Anglo-Saxon heredity. Moreover, they all called on the state to protect the Anglo-Saxon nation from the inferior blood being injected into its veins every day of every year.[89]

Anglo-Saxonism arrived at American universities in the 1880s. When Rauschenbusch imagined that, in the ancient German forests, "the social supremacy of the Aryan race manifested itself and got its evolutionary start," he was borrowing a notion popular among professional historians that the capacity for democratic government was hereditary, a race trait unique to the Anglo-Saxon people.[90]

Anglo-Saxonism dominated the views of the first professional historians in the United States, who claimed that American political institutions were the lineal descendants of ancient Saxon practices. An influential proponent was Herbert Baxter Adams (1850–1901), the prime mover behind the founding of the American Historical Association in 1884. Adams taught his Johns Hopkins graduate students that American political history should be understood as a matter of heredity. The "germs" of liberty and self-governance were Saxon in origin, transmitted first to England and then to America.[91] If Great Britain is our motherland, John Burgess explained, Germany "is the motherland of our motherland."[92]

Woodrow Wilson's *The State* (1889), a survey of the origins of government completed under Adams' supervision, began by announcing its focus would be on the governments of the "Aryan races."[93] American democracy was not an American invention. It was, Wilson wrote, a racial inheritance dating to the Teutonic tribes of ancient Germany. The English colonists in America had simply let their "race habits and instincts have natural play."[94]

Other races lacked the founding fathers' capacity for democratic government, which was a uniquely Anglo-Saxon inheritance. It was a deeply significant fact, Wilson declared, that democracy had taken root only in the United States and in a few other places "begotten of the English race." To account for democratic Switzerland, Wilson, in good Lamarckian fashion, explained that the Teutonic virtues had been transmitted by habit, not instinct. Race determined whether democracy succeeded or failed.[95]

Germ theory found its way into other corners of the newly emerging university social sciences, including psychology. G. Stanley Hall, a developmental psychologist who was a colleague of H. B. Adams at Johns Hopkins, went on to become the first president of the American Psychological Association (1892) and president of Clark University. Hall's conception of psychological development borrowed a closely related notion from evolutionary thought: recapitulation theory.

Recapitulation theory posited that an individual organism's development from embryo to mature adult recapitulates the stages of development of its evolutionary ancestors. So, just as the historians claimed that ancient Saxon ideals were recapitulated in contemporary American political institutions, so did Hall's developmental psychology claim that the ancient fish ancestors of *homo sapiens* were recapitulated in human embryos at an early stage of their development. By the same logic, human infants were less-evolved creatures than were mature adults, and children, though human, were at the level of "savages."[96]

The reverse was also held to be true. The savage races were like children, their development arrested at an evolutionary stage that the superior races had progressed well beyond. And, like children, the savage races were incapable of self-government, requiring the paternalistic protection of their betters.

8

Excluding the Unemployable

The Progressive Era catalog of inferiority was so extensive that virtually any cause could locate some threat to American racial integrity. There were degenerate Anglo-Saxon hill clans, immigrants from southern and eastern Europe and from Asia, backward peoples in the territories of the new American empire, African Americans, the feeble-minded, the epileptic, the insane, the alcoholic, the syphilitic, the tubercular, the congenital criminal, the pauper, the prostitute, the tramp, the factory girl, the "sterile" educated woman, and the shirking immigrant father who neglected to provide a family wage, among others.

Those undesirables belonging to more than one category were doubly scorned, as when the social-work journal, *The Survey*, exclaimed: "the feeble-minded woman at large is the most dangerous person the state can harbor!"[1] Simon Nelson Patten observed, with approval, the manifold ways in which Americans differentiated and distanced themselves from inferiors deemed threats:

> The South has its negro, the city has its slums, organized labor has its "scab" labor, and the temperance movement has its drunkard and saloon keeper. The friends of American institutions fear the ignorant immigrant, and the workingman dislikes the Chinese.

Every American, Patten said, wanted to differentiate himself from some "other class or classes which he wishes to restrain or exclude from society."[2]

* * * * *

What did these many inferiors have in common besides their purported threat to American heredity? They worked for wages, which in the eyes of many economic reformers made them doubly dangerous. Progressive economists led the way with a theory of how hereditary inferiority threatened both the American workingman and American racial integrity.

Before the American Civil War, inferior workers had known their place, and if they didn't, they were shown it. Slaves, coerced by state and extralegal violence, were forced to do the dirtiest and hardest work. The unskilled, many of them Irish immigrants, wielded the pickaxe and shovel. Women and domestics scrubbed, cleaned, and hauled. Apprentices and "hirelings" did what their bosses told them to do.

In antebellum America, "place" referred to social location and also to physical location. Slaves and farm hands were in the field; laborers were on the road cuts, canal ditches, and in the mines; women, domestics, and servants were in the kitchens and basements; and the other underlings were in the backs of shops. Who you were determined what work you did, and, before industrialization, where your work was.

After Emancipation and Reconstruction, where inferiors worked was increasingly centralized in the North. Industrialization drew all these workers to urban employment, along with their republican betters—the propertied farmers and journeymen—and others pulled to factories by the same centripetal force of better pay. Inferiors were now visible and were perceived to be economic competitors. They still had to know their social and political place, but increasingly they were working in the same location as their social superiors.

The progressive economists' living-standard theory of wages, we saw in Chapter 5, was a way of capturing a widespread anxiety: if capitalists can hire whomever they care to, the work will always go the lowest bidder. Insofar as labor productivity was irrelevant, there was a race to the bottom, and the cheapest labor won.

Thus did many observers accuse inferiors of accepting low wages and undercutting the American workingman. Sometimes inferior workers were portrayed as the exploited dupes of the capitalist. At other times they were portrayed as the capitalist's accomplices. Often they were made out to be both. In all events, the threat was the same: the low standards of inferior workers.

By 1910, 22 percent of the US labor force was foreign born. Women made up 21 percent of the labor force.[3] Black Americans were still trapped in the Jim Crow South, but the great migration northward was beginning. How many workers were disabled was harder to measure, but, all in all, the "inferiors" looked to be on the march.

THE MENACE OF THE UNEMPLOYABLE

The term "unemployable," popularized by Sidney and Beatrice Webb, was a misnomer, for many of the unemployable were, in fact, employed and others desperately wanted to be. The Webbs used the term to describe people incapable of work, as well as those who could work but who accepted wages below a standard reformers judged acceptable. The latter group posed the threat.

University of Chicago sociologist Charles Henderson put it plainly: the unemployable were those who "bid low against competent and self-supporting men who are trying to maintain or raise their standard of living; and they can do this just because they are irresponsible and partly parasitic."[4] By "parasite," Henderson meant that the unemployable worker earned less than was required to support him- or herself, creating a shortfall that had to be met by other members of the worker's household or by private or public charity.

Henderson borrowed "parasite" from Sidney and Beatrice Webb's *Industrial Democracy*, which was influential among American labor reformers. The Webbs affixed the term to sweatshop industries that paid wages below a living wage, and to the workers who accepted these wages. "Parasite" recalled Karl Marx's vivid characterization of capitalists as vampires, while also evoking the older pejorative, pauper, the poor person dependent on charitable assistance.

Since parasites, by assumption, could not pay their own way, their economic competition served only to drag down the wages of their betters. Letting the unemployable work was thus socially destructive, so, went the argument, they should be removed from the work force, kept at home, segregated in rural labor colonies, or placed in institutions.

Of all ways of dealing with these unfortunate parasites, the Webbs opined, "the most ruinous to the community is to allow them unrestrainedly to compete as wage earners." For the unemployable class, "unemployment is not a mark of social disease, but actually of social health."[5] When New York state established America's first industrial labor colony in 1911, it was applauded by the AALL as a scientific and humane method of keeping the unemployable out of labor markets, thus improving employment prospects for the worthy poor.[6]

Richard T. Ely hailed the virtues of segregating the unemployable in labor colonies, but when segregation was insufficient for the more intractable cases, Ely offered more drastic remedies. "The morally incurable" and those "who will not work and will not obey," Ely asserted, "should not be allowed to propagate their kind."[7] The problem went deeper than the economic competition of the disabled. The problem, Ely declared, was the very "existence of these feeble persons."[8]

Those truly unable to work were portrayed somewhat more sympathetically. But the unemployable who had the temerity to accept employment were condemned as low-wage threats to their betters.

Compared with the United States, England had relatively little immigration, and English eugenicists focused their fears on what they called the industrial residuum, the mentally and physically disabled, and those members of the working class they deemed paupers and criminals. The Webbs' classified the unemployable as

> the sick and the crippled, the idiots and lunatics, the epileptic, the blind and the deaf and dumb, the criminals and the incorrigibly idle, and all those who are actually "morally deficient" ... and [those] incapable of steady or continuous application, or who are so deficient in strength, speed or skill that they are incapable of producing their maintenance at any occupation whatsoever.[9]

In Henderson's social work text, *The Dependent, Delinquent and Degenerate Classes*, the unemployable were paupers, the feeble-minded, and persons of low moral character. Ely's unemployable group consisted of "the defective, the delinquent and the dependent."[10]

Walter Lippmann, writing in the *New Republic*, baldly asserted that workers who were "old, or weak-minded, or physically feeble, or so utterly untrained and illiterate that under American conditions they cannot be employed at a living wage," should not be permitted "to debauch the labor market, to wreck by their competition the standards of other workers." For Lippmann, employment for the unemployable was economic debauchery.[11]

By the end of the First World War, the catalog of unemployables had grown so large it invited satire. Columbia sociologist Franklin Giddings, returning in 1919 from a national meeting of American social workers, criticized as inadequate their standard classification scheme for disabled Americans—defective, dependent, and delinquent.

Tongue in cheek, Giddings proposed a more refined terminology, one that sorted the disabled into the Depraved, the Deficient, the Deranged, the Deformed, the Disorderly, the Dirty, and the Devitalized, all seven of which were to be seen as subclasses of the Defective, the Dissolute, and the Depleted.[12] The editor may have failed to notice that Giddings's piece was sarcastic. The editorial printed just a few pages before it was deadly earnest. Titled "Stopping the Undesirables," it called for extending wartime restrictions on immigration.[13]

Nobody in Progressive Era America wanted the handicapped having children, but this group did appear to be a lesser threat to American wages than were the far larger groups of interlopers, immigrants, and women. John R. Commons estimated in 1890 that "only" 2 percent of the US population was mentally or physically defective.

LOW-STANDARD WORKERS

The American discourse of labor inferiority did not invoke the quantity of workers; it was concerned with the quality of workers. The competitive threat of immigrants, women, and the disabled was not attributed to labor-supply effects on wages. The competitive threat instead was found in the putative willingness of the inferior, because of their low standards, to accept low wages. As Simon Nelson Patten observed, "the cry of race suicide has replaced the old fear of overpopulation."[14]

"Race suicide" was a Progressive Era catchphrase, coined by the captious Edward A. Ross to describe the theory that races compete, and racial competition is subject to a kind of Gresham's Law (that is, bad heredity drives out good). Workers of inferior races, because they are able to live on less than the American workingman, accept lower wages. American workers refuse to reduce their living standards to the immigrant's low level, so, in the face of lower wages, opt to have fewer children. Thus did the inferior races outbreed their biological betters.

The low-standard or undercutting-of-wages part of the theory got its start with the violent activism of white Americans against Chinese immigrant workers. The title of a pamphlet published by the American Federation of Labor trenchantly captured the heart of the claim: *Meat versus Rice: American Manhood against Asiatic Coolieism, Which Shall Survive?*[15] If wages were determined by living standards rather than by productivity, then the meat-eating Anglo-Saxon could not compete with the Chinese worker accustomed to rice.

Professor Woodrow Wilson, in his popular *History of the American People*, proffered the same theory of low-standard races undercutting American wages, adding a fillip of racism to cement the notion that race explained the low standards. White laborers, unable to "live upon a handful of rice for a pittance," could not compete with the Chinese, "who with their yellow skin and strange debasing habits of life seemed to them hardly fellow men at all but evil spirits, rather."[16]

As Ross put it, the Coolie "cannot outdo the American," but "he can underlive him."[17] American workers were more productive, Ross claimed, but because Chinese immigrants accepted lower wages, they underbid the American workingman. The Chinese took American jobs not because they were more productive, but because they worked cheaply.

These broadsides against the Chinese were published twenty years after the Chinese Exclusion Act had already banned Chinese immigrants from work in America. They were issued because the Act was up for renewal. Economic reformers wanted Chinese exclusion made permanent, and they succeeded. They also saw racial exclusion of the Chinese as a model for expanding restriction, so they redeployed the low-standard-races theory to justify the restriction of southern and eastern European races.

Immigrants earned wages in America, but they did not command American wages because they did not have American standards. Anti-immigrant reformers used "American wages" or the "American standard" to refer to the wages of white, male, Anglo-Saxon heads of household. Immigrant races, women, African Americans, and the disabled all had different—lower—standards, which made their economic competition a threat to the American standard.

John Graham Brooks, Unitarian minister and first president of the National Consumers League, put it plainly: standards of living were a "question of race." The League's white label, Brooks said, guaranteed garment consumers that their clothing was made under conditions that maintained "the white as against the cooly [sic] standard of life."[18]

The Coolie-standard indictment initially targeted the Chinese, but reformers readily applied it to other races and peoples. John R. Commons and John B. Andrews informed readers of their *Principles of Labor Legislation* that Chinese, Japanese, and Hindu immigrants willingly "accept wages which to a white man would mean starvation."[19] Commons and Andrews judged the exclusion of "Orientals," also enacted in other high-immigration countries, such as Canada, Australia, and New Zealand, to be a matter of white economic survival. Ross, notorious for his militancy against Asian immigration to the United States, once suggested that "should the wors[e] come to the worst, it would be better for us if we were to turn our guns upon every vessel bringing [Asians] to our shores rather than to permit them to land."[20]

Ely also found time to sing in the Asian-exclusion chorus. He indicted the Chinese workers for their "hard way of living," calling their propensity for low wages a threat to American workers.[21] Ely condemned the Chinese not merely for their economic competition, but also because racial difference threatened American national unity.

The fullest unfolding of our national faculties, Ely asserted, required "the exclusion of discordant elements—like, for example, the Chinese."[22] Ely assumed that a unified American nation required racial homogeneity. As for south Asians, Ely proposed that famine-relief efforts in India should be suspended. Why not, Ely ventured, "let the famine continue for the sake of race improvement?"[23]

Making wages a function of race (and of gender or intelligence) was vital to the argument for exclusion. The economic reformers who vilified the Chinese essentially accused them of being hard working, law abiding, frugal, and resourceful. These virtues were American virtues—welcome, indeed, admired in every other context. To demonize the Chinese, their critics had to render them un-American in some other fashion. They found their answer in race, and in race's putative effects on living standards and wages.

The Chinese, as Woodrow Wilson put it, had yellow skin and strange debasing habits of life. Their race and their race's low standards were what made them un-American. Crucially, race, like sex or intelligence, was immutable, and it came readymade with homegrown connotations of inferiority. There was no changing the inferior's low standards.

White men belonging to ethnic groups that had arrived in America beginning in the mid-nineteenth century—Irish, German, and Scandinavian—might grudgingly be admitted to the category of "American," but only in the service of excluding other inferior groups. Irish immigrants, for example, were regarded with fear and loathing by the Boston Yankee gentry, whose social and political authority they had threatened and gradually usurped.

In San Francisco, however, where whites made the Chinese worker the paramount economic threat, Irish immigrants were promoted to "American," so long as they joined the opposition to the Yellow Peril. Ross, ever opportunistic, modified his formulation this way, "Reilly can outdo Ah San, but Ah San can underlive Reilly."[24]

It was the same theory that low-standard races threaten American workers, but one that admitted the Irish immigrant, at least the Irish male, to the category of "American."[25] Ross promoted to "American" Irish immigrants in California, so long as they supported excluding the Chinese, but he denied the same privilege to millions of other European immigrants who found themselves in eastern and midwestern cities.

RACING TO THE BOTTOM

Profoundly illiberal, the living-standard theory of wages was also economically problematic. Why would an immigrant willingly accept wages lower than the value of his or her labor? Plausible economic explanations could be constructed. One was monopsony. The immigrant might be employed in a company town, with both wages and spending at the mercy of his or her employer. Or the immigrant, bound to a religious community, might be unwilling to accept work too far from home, or, even if mobile, might be unaware of better work opportunities.

Yet the living-standard theory posited that inferior workers accepted low wages because their race or other innate debility predisposed them to a low standard of living. Low-standard theorists sometimes slipped from the claim that low standards were innate to the claim that low standards were cultural in cause, "race habits" rather than "race instincts." But the slippage was rarely noted, and with Lamarckian inheritance, cultural standards of living could become hereditary.

The living-standard theory of wages was vague on key details. Were the inferior workers displacing the American workingman, or were they filling menial jobs created by a growing economy, jobs the American workingman would not consider taking? If Anglo-Saxon workers were more productive, as the theory invariably assumed, why had they not moved up to better jobs, their superior skill justifying higher wages?

The living-standard theory of wages also tended to confuse living standards (which are determined by spending) with wages (which are income). A frugal standard of living, whatever its origins, in no way entailed accepting low wages. Living cheap did not mean working cheap.[26] A thrifty immigrant could spend little without accepting wages less than he or she had to, saving the difference to fund a small business, to buy passage for other family members, or to send remittances to the old country.

These theoretical weaknesses proved no bar to popularity. The charge of undercutting American workers was made against nearly every immigrant group trying to gain a foothold in the factories, shops, and mines of industrial America.

Jews fleeing oppression in Russia and eastern Europe were regular targets. Commons argued that the Jewish sweatshop was "the tragic penalty paid by that ambitious race." Like Ross's Coolie, Commons's Jew was less productive but lived cheaper than the American workingman. The economic competition of the Jews thus forced American workers to have fewer children. "Competition has no respect for the superior races," Commons said; so "the race with lowest necessities displaces others."[27]

The progressive muckraker and settlement-house worker Jacob Riis was regarded as sympathetic toward working-class immigrants, whose wretched living conditions he documented in his sensational 1890 exposé, *How the Other Half Lives*. But Riis, himself an immigrant, was decidedly ambivalent about the eastern European Jews living in tenement New York, and he too used race to explain labor market outcomes. The Polish Jew, Riis explained, worshipped money and lived cheap. The "instinct of dollars and cents" was so strong in Jewish children, Riis wrote, they could count almost before they could walk. The Jews had monopolized New York's garment business by means of their low standards. Riis perfectly encapsulated the living-standard theory of wages when he asserted, the Jew's "price is not what he can get, but the lowest he can live for and underbid his neighbor." Some of the other half, Riis made clear, were not victims deserving sympathy, but threats deserving scorn.[28]

Few immigrants escaped the charge of undercutting American workers. Frederick Jackson Turner accused Italians, Slovaks, Poles, and Russian Jews of having "struck hard blows since 1880 at the standard of comfort of the American workmen." These immigrants, Turner wrote, had turned New York City into a great reservoir for the pipelines draining "the misery pools of Europe."[29]

Eugene Debs accused "the Dago" of underbidding the American workingman by living even more "like a savage or a wild beast" than did the reviled Chinese immigrant. To his list of low-standard immigrants Debs added subhuman Slavs, degraded Huns, and other "pagan labor scourges." These slum dwellers, Debs charged, had successfully invaded American shores only because they were willing to work for wages on which an American would starve, by living on "scavenger food ... in dens that an American dog would bark at."[30]

The Wharton School's radical economist Scott Nearing complained that if "an employer has a Scotchman working for him at $3 a day [and] an equally efficient Lithuanian offers to do the same work for $2 ... the work is given to the low bidder."[31] Ross as ever, offered the bluntest portrayal: "Yankee Jim does not rear as many youngsters as Tonio from the Abruzzi, because he will not huddle his family into one room, feed them macaroni off a bare board, work his wife barefoot in the field, and keep his children weeding onions instead of in school."[32]

William Z. Ripley depicted labor market competition as a race toward the racial bottom. The "American workman is underbid by the Scandinavian. He in turn is cut under by the Jew and Bohemian. The Pole will take even less than these, and finds at last his standard of living undermined by the Syrian and the Armenian."[33]

MAKING INFERIORITY LEGIBLE

The theory of undercutting inferiors faced a measurement challenge. How to identify the low-wage threats? Consider the disabled first.

As we have seen, many economic reformers called for the so-called unemployable to be removed from work. But how were they to be distinguished from other employees? Using Sidney and Beatrice Webb's catalog of the unemployable, the crippled, the blind, the deaf, and the dumb would probably eventually be found out, as might the epileptic. Just as some African Americans found ways to pass as white, the handicapped found ways to pass as employable, but their handicaps would have been hard to conceal from those who wished to deny them gainful employment.

How to identify the paupers, the alcoholics, the tramps, the congenital criminals, and the lazy? Their purported defects where less conspicuous. One answer was the time clock, which logged the worker's time of arrival and departure. American time clocks were patented in the late 1880s. The merger of several established American time clock companies in 1911 formed a company that eventually took the name of International Business Machines. The time clock identified the employees who were less punctual or less reliable.[34]

What of the mentally disabled, known as defectives and the feeble-minded? The high functioning among them, many of whom were passing as normal, presented a trickier problem for reformers warning of the unemployables' threat. As we saw in Chapter 4, one method devised for detecting them, intelligence testing, was deployed during the First World War. In the early 1910s, Harvard psychologist Hugo Münsterberg was already advising companies on how to use mental tests to screen for more efficient employees.[35]

Next, consider immigrants, who presented a somewhat different measurement challenge. The low-standard theory of wages posited that race made the immigrant an economic threat. Immigration entry stations could readily determine an immigrant's country of origin, but what needed to be known was race.

As we shall see in Chapter 9, economists and other anti-immigrant scholars devised several regulatory methods for ascertaining race and proxies for racial inferiority, such as illiteracy, poverty, and low productivity. The direct regulatory method, undertaken by the Dillingham Immigration Commission, was to develop a more scientific and more refined taxonomy of the European races, providing a kind of handbook for immigration inspectors charged with indentifying immigrant races.

A more indirect but less challenging method of detecting inferiors was the literacy test, first proposed by progressive economist Edward Bemis as a method for sorting the desirable immigrants from the northwest of Europe from those arriving from Europe's southern and eastern countries. Francis Amasa Walker, president of MIT and of the AEA, proposed a wealth test. Requiring immigrants to post a bond of $100, Walker argued, would sort the superior from inferior races.

Literacy and wealth might proxy for fitness at immigration stations, but what of inferiors already in America and in the work force? Administering literacy tests or mandating financial bonds for the millions already in America was impractical. A far more practical and more efficient method zeroed in on what all inferiors shared in common, low labor productivity. A wage test, which identified inferiority with low labor productivity, would catch *all* inferiors with low standards, the unemployable, the immigrant, and the woman.

A legal minimum wage, applied to immigrants and those already working in America, ensured that only the productive workers were employed. The economically unproductive, those whose labor was worth less than the legal minimum, would be denied entry, or, if already employed, would be idled. For economic reformers who regarded inferior workers as a threat, the minimum wage provided an invaluable service. It identified inferior workers by idling them. So identified, they could be dealt with. The unemployable would be would be removed to institutions, or to celibate labor colonies. The inferior immigrant would be removed back to the old country or to retirement. The woman, as we shall in Chapter 10, would be removed to the home, where she could meet her obligations to family and race.

9

Excluding Immigrants and the Unproductive

In October 1886, President Grover Cleveland presided over the dedication of the Statue of Liberty, also known as Liberty Enlightening the World.[1] The robed female figure of Liberty, carrying the text of the Declaration of Independence and bestriding broken shackles of oppression, bore aloft a torch to light the way of the world's oppressed peoples arriving in New York Harbor, immigrants seeking the refuge she promised to all. But Liberty, erected belatedly owing to fundraising difficulties, represented ideals that were already under legal assault.

The growing American administrative state was gradually monopolizing the authority to regulate individual rights to free movement across political boundaries.[2] In 1883, the Supreme Court had struck down the Civil Rights Act of 1875, abetting national acquiescence in the brutal reestablishment of white supremacy in the American South. A year before, the US government initiated its campaign to end a century-long era of open immigration, commencing with the Chinese Exclusion Act of 1882. This Act barred nearly all Chinese immigration and prohibited naturalization of remaining Chinese immigrants, stigmatizing them as unassimiliable racial inferiors.

That same year, the Immigration Act established a head tax on immigrants and barred entry to aliens who were suspected of being convicts, paupers, idiots, or lunatics. Under pressure from anti-Catholic nativist groups, Congress in 1885 outlawed what American governments had once encouraged—contract labor (that is, immigrants whose passage was paid by their employers).

Restrictionists wanted more and pressed for higher barriers. They got them, albeit belatedly, overcoming both the long tradition of American openness and a political coalition fighting to maintain it. After twenty years of ongoing Congressional attempts, the Immigration Act of 1917 passed the literacy test designed to exclude immigrants from southern and eastern Europe.

The Johnson-Reed Immigration Act of 1924, which made permanent the Emergency Quota Act of 1921, imposed annual national-origins quotas equal to 2 percent of each nationality's population in the United States in 1890, a year chosen to effectively terminate the immigration of Catholics, Jews, and Orthodox Christians from southern and eastern Europe. It worked. Immigration from eastern and southern Europe, which had averaged 730,000 per year in the decade before the First World War (1905–1914), plummeted to a scant 20,000 persons per year, a reduction of 97 percent.[3]

With less fanfare but with still more draconian measures, the Johnson-Reed Act also authorized a ban on all immigration from Japan and a host of other counties in a vast "Asiatic Barred Zone," which ran from Turkey eastward through the Middle East to India, southeast Asia, and Indonesia. In the quarter century following, fewer immigrants arrived in the United States than had arrived in the single year of 1907.[4] A nation peopled almost entirely by immigrants and their descendants effectively closed its gates.

Hostility to immigrants, like race prejudice, was nothing new in America. As Rogers Smith, Gary Gerstle, Desmond King, and other scholars remind us, American nativism, like American racism and sexism, was not the occasional mild fever. It was a chronic, debilitating illness.[5]

America had a long and ignominious tradition of nativist intolerance, dating to the short-lived Aliens Act of 1798, which empowered the president to arrest or deport any alien deemed dangerous. In the 1840s and 1850s, the Know Nothings of the American Party gained widespread political support by vilifying Irish immigrants fleeing famine and German immigrants fleeing revolution as un-American threats to the nation.

What did change, beginning in the 1880s, was the role of the administrative state in immigration regulation and government's use of social scientific expertise to investigate and advise on immigration policy. Legislation in 1891 established a federal Bureau of Immigration and funded construction of the Ellis Island entry station. Subsequent laws enhanced the Bureau's investigatory and regulatory powers, and added to the growing list of undesirables to be excluded.

Anarchists, polygamists, and epileptics were barred in 1903.[6] The same act required passenger manifests to record the race of every entrant. In 1906, the Bureau of Immigration added naturalization to its regulatory portfolio.

The Expatriation Act of 1907 required American women who married foreigners to surrender their US citizenship. The movement of peoples was gradually but steadily taken over by regulatory authority, with its methods of surveillance, inspection, documentation, and deportation.

With regulatory control came scientific investigation by experts on the causes and consequences of immigration, best represented by the vast investigations of the Dillingham Commission (1907–1910), chartered by Congress to survey the "immigrant problem." The largest federal investigation yet undertaken outside the Census, the Commission employed a staff of 300 field agents, statisticians, and clerks; it surveyed more than 3 million immigrants in 300 communities.[7] Its experts produced forty-one volumes, weighing in at nearly 29,000 pages, investigating immigrant households, schools, banks, charity seeking, criminality, head shape, and much more. First and foremost, however, the Dillingham Commission investigated how immigrant races adversely affected American workers' wages and employment, dedicating half of its volumes to immigrants in American industry.[8]

Progressive economists were at the forefront of the scholars and activists joined in an anti-immigrant campaign begun not long after the American Economic Association (AEA) was founded in 1885. In 1888, the AEA offered a prize for the best essay on the evils of unrestricted immigration.[9] A few months after the Statue of Liberty was lit, progressive economist Edward Bemis devised the literacy test as a technique for identifying and "rigorously excluding the plainly unfit."[10]

In principle, a literacy test did not discriminate by race or nationality. But Bemis devised it precisely because, in practice, it would eliminate the bulk of what he termed the "new immigration," which in 1888 largely comprised Italians, Poles, and Hungarians, while still admitting the desirable races from northwestern Europe, the "old immigration" of Swedes, Germans, English, Scotch, and most of the Irish.[11] The literacy test's great virtue, Bemis said, was that it identified and excluded inferior immigrant races.

When Bemis first proposed excluding them, immigrants from eastern and southern Europe made up only one-quarter of total immigration. While total immigration declined in the depression-plagued 1890s, anti-immigrant sentiment grew with the share of "new" immigrants, which doubled to 50 percent by 1900, and averaged 70 percent in the first decade of the twentieth century.[12] Adopted by the Immigration Restriction League, founded in 1894, Bemis's literacy test became the policy centerpiece of anti-immigration agitation.

RACE SUICIDE

Progressive scholars lent vital intellectual support to the Progressive Era anti-immigration campaign. Richard T. Ely, proud of his former student, was quick to applaud the selective virtues of Bemis's literacy test.[13] In the same year, Columbia economist Richmond Mayo-Smith published a series of influential articles defending immigration restriction, which he gathered in his 1890 book, *Emigration and Immigration*.[14] Mayo-Smith was a cofounder of the AEA and was later elected to the National Academy of Sciences. Edwin R. A. Seligman described him as "indisputably the foremost American scientific statistician" of his time.[15]

The American tradition of providing asylum, Mayo-Smith said, should not apply to the paupers, convicts, and cripples streaming into America. The first European settlers of America rightfully seized possession of it from "a few thousand savages," because the settlers were of higher civilization, and progress required that the superior be granted "the moral right to triumph."[16] The more recent European immigrants, Mayo-Smith argued, could claim no such right, because they were inferiors.

Beginning with his presidential address to the AEA in 1890, Francis Amasa Walker offered a race-suicide account that proved especially influential in the immigration debate.[17] Walker, a decorated Civil War officer, president of MIT, and two-time superintendent of the US Census, lent his name and scientific reputation to a theory that joined fertility to the low-wage threat of immigrants. Walker argued that the superior American workingman could not compete with the low-standard immigrant races. Rather than reduce his living standards to the immigrant's level, the American worker chose instead to have fewer children.

Walker disparaged the newcomers from southern Italy, Hungary, Austria, and Russia as "beaten men from beaten races, representing the worst failures in the struggle for existence." Their putative inferiority notwithstanding, the newcomers' low standards permitted them to undercut and outbreed their biological betters, displacing the "native" Americans. Without immigration barriers, Walker warned, "every foul and stagnant pool" of Europe would soon be "decanted upon our shores."[18]

Walker, like Mayo-Smith, was one of the United States' most eminent scholars of population, but his immigration-reduces-native-fertility theory stood on a shaky foundation.[19] Walker observed that early nineteenth-century forecasts of the US population in 1840 and 1850 assumed little or no immigration, but nonetheless proved quite close to the actual figures. Since immigration had, in fact, added 2.3 million people in the 1830s and 1840s without changing expected population growth, Walker concluded immigration itself must have caused an equal decline in native births.[20] Immigrants, then, did not add to the native population, but displaced it. And because, Walker maintained, immigrants from southern and eastern Europe were inferior, they endangered American racial health.

Walker's theory, unburdened by direct evidence as it was, later attracted some critics, who argued that the decline in native fertility preceded the increase in immigration and was caused not by the menace of cheap labor, but by increased urbanization, higher living standards, and later age of marriage.[21] Walker, however, had won honorary degrees from Harvard, Yale, Columbia, St. Andrews, Dublin, Halle, and elsewhere, and his scientific reputation carried the day.

Walker preferred a wealth test to the literacy test. Requiring immigrants to post a bond of $100, Walker argued, would eliminate nearly all the undesirable races. Very few Italians, Poles, or Jews would have a sum this large, whereas the bond would still admit "thrifty Swedes, Norwegians, [and] Germans." A wealth test also kept out literate anarchists, criminals, and drunkards. It was difficult to evade and more efficient than administering tests in fifty languages.[22]

Walker died 1897, never witnessing the full legislative flowering of his race suicide theory. But anti-immigrant scholars of all stripes appealed to his authority over the many years of Progressive Era restriction agitation. For example, social worker Robert Hunter warned of race suicide in his influential book *Poverty* (1904). The most serious threat from immigrants, Hunter wrote, was not their tendency to pauperism and criminality but their higher fertility. More immigrant children threatened the "annihilation of the native American stock," and with it American freedom, American religion, and American standards of living.[23]

Henry Pratt Fairchild, Yale economist and author of *The Melting Pot Mistake*, appealed to Walker's displacement theory when he maligned immigrants as "supplanters of native children."[24] Andrew Dickson White, president of Cornell University, portrayed immigrants as barbarian invaders who threatened to overwhelm American civilization just as Huns and Goths had sacked Rome.[25] Prescott Hall, cofounder and standard-bearer of the

Immigration Restriction League, concocted an even more lurid metaphor. Native children, he said, were being "murdered by never being allowed to come into existence, as surely as if put to death in some older invasion of the Huns and Vandals."[26] Suicide, annihilation, displacement, invasion, and murder—this was the language of American scholars warning of race suicide.[27]

* * * * *

The rise of race-suicide theory illustrates how evolutionary concepts, once employed to defend immigration, could now be used to attack it. Before 1880, European immigrants to the United States were regularly valorized as naturally selected. Darwin, in the *Descent of Man*, attributed the progress and character of the American people to the results of natural selection. Ten or twelve generations of Europe's most energetic and restless people, Darwin said, emigrated to "that great country and have there succeeded best."[28] Enterprising people of good stock left the Old World, and the rigors of American frontier life selected the fittest among them.

The Darwinist idea of America as a crucible for producing a vigorous American racial character was a recurring motif in the work of historian Frederick Jackson Turner, Ross, and other scholars enamored of the American frontier.[29] And yet, when the sources of immigration to the United States changed so that immigrant peoples were increasingly Catholics, Jews, and Orthodox Christians from southern and eastern Europe, scholars turned Darwin on his head, depicting the immigrants not as selected and hence fitter, but as unfit.

Mayo-Smith awarded the honorific of "colonist" to the early Anglo-Saxon settlers. The more recent European arrivals were mere "immigrants." When, in 1888, he lamented that nearly half of the white population was foreign born or descended from the foreign-born, Mayo-Smith forgot that the other half of the white population, the Anglo-Saxon stock, was itself descended from the foreign-born.

The distinction persisted. Commons condemned the more recent arrivals as inferior to those of the seventeenth and eighteenth centuries. Only one race had been fit enough to plunge into the wilderness, battle Indians, and establish frontier farms.[30] Social worker and economist Edward T. Devine, writing in *The Survey*, claimed that colonial era conditions selected only the exceptional. The immigrant of 1903, however, followed a path made easy and thus avoided the rigors of selection.[31] Once celebrated as vigorous and enterprising, immigrants to America now were condemned as weak and opportunistic. Natural selection, anti-immigrant scholars asserted, had ceased operating. But plenty of evolutionary explanations remained at hand.

Ross cited the closing of the American frontier announced with the 1890 Census. "Anthropologically we are at a zenith," he lamented, "for the westward shifting [of] people has slackened and the tonic selections of the frontier have well-nigh ceased."[32] Commons pointed to the vice-ridden and insalubrious environment of American industrial cities as selecting for the unfit. Theodore Roosevelt blamed the established Anglo-Saxon elite for failing in its racial duty to produce sufficiently large numbers of offspring.

Commons added to the restriction chorus with his study of the economic effects of immigration conducted in 1900 as a special agent for the US Industrial Commission (USIC). Commons spent months investigating garment industry sweatshops. Whether an immigrant belonged to an inferior or merely backward race, he or she brought a lower standard of living, which undercut American wages.

Commons's remedy was to organize labor. By keeping wages high, unions would reduce the adverse economic and racial consequences of wage competition. The problem, Commons wrote, was the immigrant would not be organized. He called it "the menace of immigration to labor organization," a threat also identified by Samuel Gompers of the AFL, with whom Commons collaborated when they both worked at the National Civic Foundation in 1902.[33]

Commons's USIC report singled out the Jews in particular. The problem was the Jew's "commercial instinct," which led him to join a union only when it offered a bargain, abandoning the union as soon he got or failed to get what he wanted.[34] Between the turn of the century and the First World War, Jews joined unions in large numbers, and some became leaders of American labor organizations, making Commons's accusation look preposterous. Forced by events to retreat, Commons gracelessly conceded, "most remarkable of all, the individualistic Jew from Russia, contrary to his race instinct, is joining the unions."[35]

But Commons did not abandon his anti-immigrant stance, finding a natural home with Prescott Hall's Immigration Restriction League. When he reviewed Hall's exclusionist tract, *Immigration and Its Effects upon the United States,* he acclaimed the controversial book as "the most important contribution that has been made to the study of this most important American problem."[36] Commons testified before Congress on behalf of the Immigration Restriction League's literacy test in 1902, 1903, 1905, 1909, 1911, and 1913.[37]

Common's University of Wisconsin colleague Edward A. Ross was another pillar of immigration restriction among progressive economists. In his many years of anti-immigrant agitation, Ross offered his readers a surpassingly crude portrait of immigrant peoples, whom he pictured with "sugar-loaf heads, moon faces, slit mouths, lantern-jaws and goose-bill noses." The immigrants' ugliness, Ross concluded, unmistakably proclaimed their inferiority, and threatened to despoil American good looks.[38]

He maligned the new arrivals as "cheap stucco manikins from Southeastern Europe," "masses of fecund but beaten humanity from the hovels of far Lombardy and Galicia," the "slime at the bottom of our foreignized cities," the "Slavs immune to certain kinds of dirt," who brought to America a "rancid bit of the Old World," the "hirsute, low-browed, big-faced persons" who clearly "belong in skins, in wattled huts at the close of the Great Ice Age," the childish, frivolous and "cheaply gotten up *mañana* races," the "stupid and inert peoples" poaching on "the preserve of the bright and industrious," "the dullard races ... last to abandon the blind fecundity which characterizes the animal," and the "transients with their pigsty mode of life."

The blood now being injected into American veins, Ross hardly needed to conclude, was "sub-common." But among the immigrants' countless shortcomings—their ugliness, their stupidity, their servility, their politics, their bestial fecundity—the biggest threat was they worked cheap. And immigrants worked cheap, Ross asserted, because living standards were "a function of *race*."[39]

Other eminent reformers added their voices to the exclusionist cause and to the race-suicide chorus. Irving Fisher saw in rising anti-immigrant sentiment "a golden opportunity" to promote eugenics. Fisher allowed that if one set aside the question of race and eugenics, he would, as an economist, favor unrestricted immigration. But the core of immigration problem, Fisher warned, was race and eugenics, the threat of racial and mental inferiors to Anglo-Saxon race integrity.[40]

University of Chicago chaplain and sociologist Charles R. Henderson took the same tack, protesting that he did not oppose immigration. What he opposed were immigrants incapable of living "a civilized scale of life."[41] It was a matter of weeding out low-standard immigrants to protect American wages.[42]

In his popular 1910 sociology textbook, sociologist Charles Elwood advocated "selection on a large scale" to determine who shall be the parents of future Americans. The literacy test was a good start, but to it should be added a wealth test, and a biological test. A more rigorous inspection regime was needed to eliminate the "degenerate strains of the oppressed peoples of Southern and Eastern Europe."[43]

MEASURING RACES AND IMMIGRANTS

The Dillingham Commission was chartered in 1907, the year in which immigration peaked at 1.3 million, and President Roosevelt announced that race suicide was the greatest problem of civilization.[44] When it delivered its monumental report to Congress in December 1910, the Commission added its political and scientific authority to the anti-immigrant cause.[45]

Economists W. Jett Lauck and Jeremiah Jenks proved especially influential in the Commission's work. Jenks, the Cornell professor who had just completed his term as president of the AEA, shaped and supervised the Commission's work, and he deputized his protégé Lauck to oversee the Commission's studies of the economic consequences of immigration, which formed the bulk of the work. Jenks and Lauck cowrote the summary of the Commission's sprawling and voluminous investigations, in which they distilled the 29,000 pages into a brief précis, making Bemis's literacy test their centerpiece.

Jenks and Lauck represented themselves as objective scientists, asserting that they had formed no opinion on immigration restriction until the Commission's work was complete. We are not, they declared, "advocates, but interpreters of facts."[46] Protestations notwithstanding, Jenks and Lauck *were* advocates of immigration restriction, and their interpretations were littered with dubious claims about race and its adverse consequences for America.

In fact, the Dillingham Commission inquiry began by assuming what it would purport to show, that the so-called new immigrants—which it defined as immigrants from Austria-Hungary, Bulgaria, Greece, Italy, Montenegro, Poland, Portugal, Romania, Russia, Serbia, Spain, and Turkey—were far less intelligent than the old immigrants, as well as less skilled, less literate, less progressive, and less assimilable.[47] Consequently, the Commission paid scant attention to the foreign-born from traditional European sources, "British, German and other peoples."[48]

Addressing the National Conference of Charities and Corrections in 1909, Jenks explained his conception of the Dillingham Commission's mandate. First, the experts were to ascertain whether certain immigrant races were inferior to other races, and second, they were to establish whether some races were less well fitted for American citizenship than others.[49] It was time, Jenks said, to base immigration restriction on "racial characteristics" rather than individual attributes.

Jenks admitted to a certain race pride. Most of us, Jenks announced to the assembled social workers, were proud "of being Anglo-Saxons."[50] When the Dillingham Commission demobilized, Jenks churned out pamphlets on the evils of immigration for the Immigration Restriction League.[51]

Race at once motivated and bedeviled the Dillingham Commission experts studying immigration.[52] Country of origin was readily determined, but an immigrant from Austria-Hungary or Russia, sprawling multiethnic empires, could be Teutonic, Slavic, Semitic, or "even Mongolian."[53] A Czech immigrant was neither Austrian nor Hungarian, and Polish or Jewish immigrants were not Russian. A Sicilian came from the same country as a Genoan but was invariably classified as from a different and inferior race.

Immigration inspectors recorded country of origin, but the experts needed to know race. The experts said race determined whether the immigrant could be assimilated to the economic, political, and social demands of life in the American republic. But having located race as at the very center of their concerns, the experts could not easily say what it was. Turn-of-the-century race science was a muddle. Even its advocates recognized significant disagreement among authorities.

Leading race science texts, such as William Z. Ripley's *Races of Europe*, offered different and conflicting taxonomies of European peoples, founded them on different bases of classification—head shape; pigmentation of skin, hair and eyes; stature; "character"; language; religion; and so forth—and proposed different schemes for fitting European peoples into broader racial categories.

Race scholars called the English a race.[54] But they were unclear on whether the English race was coextensive with the favored Anglo-Saxon race or included other elements.[55] And, if the English race were mixed—made up of Angles, Saxons, Scandinavians, Britons, Normans, and more—what made it a race at all, and what was the status of its constituent elements?

Classification problems abounded. Ripley, for one, was unsure of what to do with the European Jews. Were they Teutonic, Alpine, Mediterranean or none of the above? He could not bring himself to admit the Irish into the highest ranks, so he made them members of the Alpine race. Even the favored race, which race scholars invariably placed at the top of their pyramids of humanity, was given different names. Ripley's Teutons overlapped ambiguously with Joseph Deniker's Nordics and Prescott Hall's Baltics.

The *Dictionary of Races and Peoples* was the product of the Dillingham Commission's efforts to authoritatively address these conundrums of human taxonomy.[56] Written by the staff anthropologist, Daniel Folkmar (1861–1932), *The Dictionary of Races and Peoples* took up the problem. He struggled mightily, if unsuccessfully, to reconcile the very different distinctions made by race science authorities.

Folkmar divided humanity into five races, Caucasian (white), Mongolian (yellow), Malay (brown), Ethiopian (black), and American Indian (red). The Malay, Ethiopian, and American Indian races did not merit further refinement. Folkmar parsed the Caucasian race into ever-finer divisions.

Caucasians could be (listed in descending order) Aryan, Semitic, Caucasic, or Euskaric. Aryans, in turn, Folkmar divided into (in descending order) Teutonic, Lettic, Celtic, Slavonic, Illyric, Armenic, Italic, Hellenic, and Iranic. These various Aryan subgroups were further subdivided into thirty-six "peoples," from Welsh to Portuguese to Lithuanian. Despite the protests of American Jewish organizations, Jews were classified as a race, designated "Hebrew."[57]

The *Dictionary of Races and Peoples*, which ran from Abyssinian to Zyrian, offered a motley compendium of ethnic stereotypes, skin complexion, head shape, and other hardy perennials of the race science literature, with some material reprinted directly from Ripley's *Races of Europe*. Bohemians, readers were informed, had a heavier brain than any other European people. The southern Italian was "excitable, impulsive, highly imaginative," unable to adapt to organized society, and, like the Corsican, prone to vendettas. The "Slav" tended to "periods of besotted drunkenness" and "unexpected cruelty." Germans in the Tyrol had a cephalic index of 87, matched only by the very broadheaded Herzegovinians. Negroes were dismissed as "the lowest division of mankind from an evolutionary standpoint," so low that Folkmar recommended against elaborating further divisions of the Ethiopian race. For the purpose of American immigration policy, all blacks were alike.[58]

* * * * *

Like Folkmar, Jenks and Lauck employed a heavily racialized rhetoric on immigration. But they used somewhat less contemptuous language to describe immigrants targeted for exclusion than did some other restrictionists crying race suicide. Theirs was not the open contempt and hostility of an Edward A. Ross, for example. In fact, they devised one justification for exclusion that, avowedly at least, did not depend upon race inferiority.

In *The Immigration Problem*, Jenks and Lauck insisted that excluding immigrants on racial grounds carried "no implication of inferiority." It only recognized, they maintained, "a difference in races and a lack of readiness to assimilate."[59] Never mind that racial capacity to assimilate was not well defined and was not supported by evidence.

To this claim Jenks added another assertion: people of any race were prejudiced.[60] Whatever science taught, all people considered their own race superior to others. It was in this context Jenks admitted to his pride in being an Anglo-Saxon.

Universal race prejudice, in turn, implied race conflict, which was intolerable, as was the alternative, wherein one race "assumes the position of a subject caste," the fate of African Americans in the South and American Indians in the West.[61] Therefore, Jenks and Lauck concluded, race-based immigration restriction was necessary to forestall additional race conflict:

> This country wants no other race problem. The negro problem is enough. Many fear that a Jewish problem threatens for a different reason. They wish to take no risks of a Chinese or Japanese or Hindu racial problem. The feeling is rather one of fear and prudence rather than one of hostility or contempt.[62]

Of course, nothing prevented anti-immigrant allies like Edward A. Ross from deploying both arguments in support of exclusion. Ross cried race suicide, while also warning of race conflict. America might absorb "thirty or forty thousand Hebrews from Eastern Europe," Ross said in 1914. But, he warned darkly, "when they come in two or three or even four times as fast, the lump outgrows the leaven, and there will be trouble."[63]

Jenks and Lauck were hardly race egalitarians. But their argument for race-based restriction on grounds of prudence rather than inferiority was taken up by a number of more egalitarian progressives, notably John Dewey.

Dewey was a genuine race egalitarian. He believed the concept of race was largely a fiction and rejected any notion of a race hierarchy. But, during the early 1920s debate over quotas, Dewey defended race-based immigration restriction on prudential grounds. Ordinary Americans, possessing primitive instincts of race prejudice, were not cultivated enough to adopt more enlightened views. Race-based immigration restrictions, Dewey argued, would reduce irrationally motivated but no less real race conflict among groups. "The world is not sufficiently civilized," Dewey concluded, "to permit close contact of peoples of widely different cultures without deplorable consequences."[64]

Another progressive with egalitarian credentials, Florence Kelley, also advocated race-based restriction. Long a resident in immigrant communities, Kelley had once opposed restriction. But a weeklong encounter with the anarchists of the International Workers of the World, who were advocating violent revolution, turned her against open immigration. Anarchy, she wrote to John Graham Brooks in 1913, "has become hereditary ... among the immigrants and their children." The Wobblies had made Kelley, she admitted, into "an active restrictionist."

Kelley proposed to exclude more than anarchists, who had been barred for a decade. "I am convinced," she confided to Brooks, "that the Pacific Coast people are right about the Mongolians; and I am sure that we are utter fools to endure the ruin of the Atlantic Coast by the invasion of Asia Minor and South Eastern Europe."[65] America, Kelley said, should *select*. Admit religious refugees but bar anarchists, the Chinese, and Asian and European peoples ruining America.

In 1913, Kelley had lived in immigrant settlement-houses for twenty-one years. In 1909, she had been a founding member of the National Association for the Advancement of Colored People. Her egalitarian credentials, she told Brooks, would safeguard her advocacy of selective restriction from the charge of race or religious prejudice.[66]

Selig Perlman was less well known than Dewey or Kelley, both progressive icons. He was one of the group of young labor economists assigned by Commons to work on the Wisconsin School's monumental *History of Labor in the United States.* Perlman described the Chinese Exclusion Act of 1882 as "the single most important factor in the history of American labour." Had the United States not excluded the Chinese, Perlman wrote, "the entire country might have been overrun by Mongolian labour, and the labour movement might have become a conflict of races instead of one of classes."[67]

Dewey, Kelley, and Perlman all advocated some form of race-based immigration restriction. They avoided race-suicide talk. But fearful of race conflict or anarchy, these progressives leaders found themselves advocating the same solution as their more extreme compatriots: immigration restriction based on race.

PROGRESSIVES AGAINST RESTRICTION

Invoking the Dillingham Commission's monumental fact gathering, Jenks liked to say that the scientific evidence for race-based restriction was incontrovertible. Jenks went so far as to claim that all field agents working for the Dillingham Commission, more than one hundred in number, became convinced that race-based immigration restriction was necessary. The experts, said Jenks, were unanimous.[68]

But, as Jenks well knew, they were not. In fact, one volume of the Dillingham report itself, *Changes in Bodily Form of Descendants of Immigrants*, cast serious doubt on a fundamental premise of race-based restriction: the fixity of racial differences. Franz Boas (1858–1942), a pioneer of physical anthropology and professor at Columbia University, found that head shape, thought to be one of the most stable and permanent characteristics of race, was instead variable. There was, Boas said, "a great plasticity of human type."

Working at Ellis Island and in New York City schools, Boas and a dozen research assistants measured the heads of thousands of immigrants, chiefly Jews, Czechs, Sicilians, and Neapolitans, along with their children. Boas found the cephalic index (the ratio of head width to head length) of American-born children differed significantly from their foreign-born siblings. Moreover, the difference between American-born children and their foreign-born siblings was increasing with the number of years their parents had lived in the United States. Whatever explained the difference between American-born and foreign-born siblings, it could not be heredity.[69]

The American-born children of eastern European Jews had longer heads than their round-headed parents and foreign-born siblings. The American-born children of southern Italians had rounder heads than their long-headed parents and foreign-born siblings. In Europe, Boas concluded drily, the two races were quite distinct, but "their descendants born in America are very much alike."[70] Using the race scientists' preferred tools, Boas challenged not only the superiority but also the very existence of the Anglo-Saxon race.[71]

This bold stance made Boas a rarity, but he was not altogether alone. A small number of economists and other scholars opposed restriction. They welcomed immigrants for the value they added to American economy and out of humanitarian concern. An even smaller minority supported immigration as a valuable source of cultural pluralism, a new concept of American nationality popularized by philosopher Horace Kallen.[72]

Immigration politics, then as today, made for strange bedfellows. The Progressive Era restrictionist coalition comprised blue-blood New England conservatives like Henry Cabot Lodge and the Harvard Brahmins who founded the Immigration Restriction League, progressive social scientists like Bemis, Commons, Ely, Jenks, Lauck, and Ross, and some organized labor leaders, such as Terrence Powderly of the Knights of Labor and Gompers of the American Federation of Labor. Gompers, whose American Federation of Labor membership reached 1.7 million in 1904, regarded the "new" immigrants as a grave threat not only to American workers' wages but also to American racial integrity.

The pro-immigration coalition was no less diverse. The arch-restrictionist Prescott Hall, when asked by the US Immigration Commission of 1899 whether there existed any scholars who opposed restriction, identified only two, Edward Atkinson and David A. Wells.[73] Both men were free-trade economists of the pre-Progressive Era generation.

Atkinson, a successful Boston textile manufacturer, saw the immigrants working in his mills as valuable additions to the workforce and to American life more broadly. Atkinson argued the United States had almost "incalculable room for immigrants," and he condemned restrictionists as "almost pusillanimous" for their refusal to offer "a refuge to the oppressed and the industrious and capable."[74]

Joining the aged laissez-faire economists, who belonged to the pre-Progressive Era generation, were business groups lobbying for a reliable supply of low-cost labor, such as the National Association of Manufacturers, and progressives who sought to preserve rather than dismantle the American tradition of providing asylum for peoples fleeing religious and political persecution, exemplified by Grace Abbott, the settlement worker who headed the Immigrants Protective League.[75]

Unlike the progressives crying race suicide, progressives with more egalitarian values were pulled in opposite directions on the exclusion question. They ordinarily eschewed race-suicide rhetoric. But some nonetheless worried about those same immigrants' effects on American workers and American national unity; others, as we have seen, advocated race-based restriction even while avowing the fundamental equality of immigrant peoples.

Emily Greene Balch, a social worker and professor of economics at Wellesley College, exemplified the more egalitarian progressive. Her extensive and sympathetic study, *Our Slavic Fellow Citizens*, published just before the Dillingham Commission presented its report to Congress, denied that the new immigrants were racially inferior. Balch also argued that immigration did not reduce American wages. The real wages of unskilled labor, she noted, had been rising from 1840 to 1900.[76] Balch also disputed Commons's claim that immigrants undermined union organization. On the contrary, she said, Slavic immigrants supported organized labor's aims, and they congregated in industries, notably mining, where unions were strong.

In forgoing race-suicide talk, Balch was an outlier among progressive economists studying immigration. Her work sympathetically portrayed the Slavic immigrants her fellow scholars so often vilified, and she offered a precociously inclusive vision of American life. Viewed from the twenty-first century, her opinions on race and nationality are decades before their time.[77]

However, even Balch was not immune to the eugenic enthusiasms of her day. While she rejected the vulgar racism of the majority of her colleagues, she did not dispute the eugenicists' premise that some groups were hereditarily inferior. She worried, for example, that if talented professional women had fewer children, "it would withdraw from the race the inheritance of some degree of picked intellectual ability," risking a serious loss of genetic quality.[78]

Balch also opposed, on eugenic grounds, subsidizing books and lunches for poor school children, warning, "if you simply want to have more people ... depraved people quite as well as any other class," then "feeding school children [is] a good thing; but if you believe it is important ... to have more

of the right kind of people, then any measure of encouragement should be most carefully selective in character."[79]

Balch's eugenic sentiments remind us that Progressive Era America was so steeped in the discourses of heredity and human hierarchy that even an egalitarian progressive—one who eschewed the racism of her intellectual circle and who courageously opposed American entry into the First World War at the cost of her professorship—nonetheless felt compelled to warn about the threat of the wrong kind of people.

* * * * *

The restrictionists failed to enact the literacy test until 1917, but only by the slimmest of margins. Congress passed a literacy-test law by a large margin in 1897, and it did so again in 1913 and 1915. It took four vetoes by three different presidents—and the hideous intervention of the First World War, which radically reduced the transatlantic movement of peoples—to keep the door ajar from 1897 and 1917.[80] In the intervening twenty years, 17 million immigrants arrived in a country of 75 million in 1900, swelling the pro-immigration constituency.

Professor Woodrow Wilson, like many scholars around the turn of the century, had sounded the alarm about the new immigration, lending credibility to anti-immigrant agitators. The last volume of Wilson's *History of the American People*, published in 1902, disparaged "the men of the lowest class from the south of Italy and men of the meaner sort out of Hungary and Poland." This class of immigrants, Wilson added, was inferior to the "sturdy stocks of the north of Europe," which arrived before 1880. The southern and eastern Europeans had "neither skill nor energy nor any initiative of quick intelligence," and their mother countries were "disburdening themselves of the more sordid and hapless elements of their population."[81]

Wilson the politician paid a political price for these sentiments when his rivals publicized them during the 1912 presidential campaign. As historian John Higham observed, "Wilson labored throughout the campaign under the embarrassing handicap of having to repudiate over and over again the contemptuous phrases he had written about southern and eastern European immigrants."[82]

The restrictionists deputized Jeremiah Jenks and Edward A. Ross to take the case for the literacy test directly to the new occupant of White House as of March 1913. Ross, who knew Wilson from his graduate school days at Johns Hopkins, sent the president some preparatory homework: Ross's anti-immigrant articles from *The Century*, which he would soon gather into *The Old World in the New*.

Wilson was unmoved by the scholarly delegation, or at least not moved enough. He vetoed the literacy test bill that crossed his desk in January 1915. It is unknown whether Wilson changed his views of the Italians, Hungarians, and Poles who came to America, or merely found it politically expedient to claim he had. Either way, Wilson's repudiation of his earlier views measured the growing political clout of immigrant groups.

His mission failed, Ross blamed the Jews, whom he vilified as clannish, shrewd, pushy, ill-mannered, underhanded, and possessed of a "monstrous love of gain." Hebrew money, Ross wrote indignantly, was financing the anti-restriction campaign, which pretended to benefit all immigrants, but was, in fact, "waged by and for one race." According to Ross, the Jews had repaid the gift of American asylum by undermining America's capacity to control its own racial destiny.[83]

A MINIMUM WAGE TO SELECT THE FITTEST

When the Dillingham Commission delivered its report to Congress, Paul U. Kellogg's *The Survey* immediately endorsed its recommendation of immigration restriction in general and the literacy test in particular. Kellogg published a dissent from Grace Abbott, who defended free immigration. But *The Survey* was vigorously restrictionist, and its editorial endorsement well captured the essential tension created by regarding the poor as victims deserving uplift but also threats requiring restraint.

Edward T. Devine wrote the editorial.[84] Devine was an economist at Columbia University, former head of the Charity Organization Society of New York, director of the New York School of Philanthropy, and past editor of *The Survey*. He was, in the words of Simon Nelson Patten's biographer, "the nation's leading philanthropic executive."[85]

In defending immigration restriction, Devine invoked a "sacred duty" to protect the American national heritage, which, Devine said, "its creators gave it to us with their blood." The crossing of races, Devine wrote, was beneficial within limits, but America had reached the dangerous point beyond which,

the biologists confirmed, interbreeding led only to a "mongrel and degenerate breed." Devine denied any "shred of bigotry or prejudice." It was not the immigrants' fault they were less skillful, less intelligent, less efficient, and less inherently desirable than native Anglo-Saxon workers. The blame lay with American employers exploiting the immigrants' inferiority, which was manifest in their willingness to accept low wages, work long hours, and remain unorganized. Admitting more immigrants, Devine wrote, amounted to treason.[86]

Paul Kellogg also strongly supported restriction, and he proposed a novel alternative: a tariff on immigrant labor. Ely had broached the idea of taxing immigrant labor to protect domestic workers as early as the late 1880s.[87] Ross had complained in 1900 that the tariff "kept out pauper-made goods, but let in the pauper."[88] Kellogg's idea was not new, but perhaps the time was now right.

Kellogg's proposal—compel all immigrants to earn at least $2.50 per day or else be denied entry—was not a tariff; it was a minimum wage. But calling it a tariff was a brilliant rhetorical stroke. The United States protected American industry with a tariff on imported goods. By the same logic, Kellogg argued, it should protect American workers by taxing imported labor.

By pushing firms to hire only the most able immigrant workers, a mandated minimum wage for immigrants would reduce the quantity of immigrants and also select for higher quality immigrants. Kellogg's minimum was nearly fifty percent higher than what the average lower-skilled worker earned in 1910.[89] One *Survey* correspondent estimated that a $2.50 per day minimum would essentially terminate the immigration of unskilled workers, reducing their numbers from 500,000 to perhaps 5,000 per year.

It was no surprise that Kellogg turned to the minimum wage. A minimum wage was the holy grail of American progressive labor reform, and a Who's Who of progressive economists and their reform allies championed it. In 1911, progressives were on the cusp of a string of legislative victories, resulting in minimum wage laws in fifteen states and the District of Columbia and Puerto Rico, beginning with Massachusetts in 1912.

As Kellogg's proposal to restrict immigration made clear, progressive labor reformers embraced the minimum wage for its power to exclude as well as to uplift. The minimum wage test would, more efficiently than the literacy test, target the inferior races of southern and eastern Europe by identifying inferiority not with illiteracy but with low labor productivity—the inability to command a minimum wage. Kellogg's race hierarchy could not have been plainer. A minimum wage for immigrants, he said, would "exclude [Angelo] Lucca and [Alexis] Spivak and other 'greeners' from our congregate industries," reserving American jobs for "John Smith and Michael Murphy and Carl Sneider."[90]

American economists engaged in the minimum wage debate of the 1910s, whether pro or con or in between, agreed that a successful legal minimum would idle the least productive workers. If the law raised the cost of hiring unskilled labor, fewer unskilled workers would be employed.

Some minimum wage advocates, such as Sidney Webb and John A. Ryan (author of the Minnesota minimum wage law), claimed that firms' labor costs would not increase, because higher wages would make workers become more productive, a view Frank Taussig called the "steam-engine theory of wages." Just as more power was obtained by putting more fuel under the boiler, so too was more labor power extracted by putting more wages into the pockets of human beings.[91] But such efficiency-wage claims—the idea that wages were more the cause than the consequence of labor productivity—were halfhearted and exceptional. Indeed, both Webb and Ryan acknowledged that a minimum wage would cause some workers to lose their jobs, namely, those whose services were worth less than the minimum rate.[92]

The theory that minimum wages discharged the least productive workers had been a constant of Anglophone political economy, dating to John Stuart Mill's (1848) *Principles of Political Economy*.[93] When England established a minimum wage with the Trade Boards Act in 1909, it did so notwithstanding the objections of a generation of England's most eminent economists—Henry Sidgwick, Alfred Marshall, Philip Wicksteed, and A.C. Pigou—all of whom observed that while the law could make it criminal to pay a worker less than the minimum, it could not compel firms to hire someone at that rate.[94] Even the intellectual champions of the English minimum wage conceded the point.[95]

The American economists agreed. A binding minimum would raise the income of some, but only by also throwing the least skilled workers out of work. Moreover, the state would have to care for the workers idled by minimum wages.

The more conservative American economists, such as J. Lawrence Laughlin, Arthur T. Hadley, and Frank Fetter, opposed the minimum wage on these grounds. Right progressives, such as John Bates Clark, were better disposed toward the minimum wage. Clark argued a minimum wage was justified when worker were paid wages less than the value of their contribution to output, but he, too, worried about the social cost of idling the least productive.

The many left progressives who advocated the minimum wage, among them Father John Ryan, Charles Henderson, Matthew B. Hammond, Henry A. Millis, Henry R. Seager, Arthur T. Holcombe, and Albert B. Wolfe, agreed that the minimum wage would throw the least productive employees out work or prevent their employment in the first place. But these reformers saw the removal of the less productive not as a cost of the minimum wage but as positive benefit to society. Removing the inferior from work was not a regrettable outcome, justified by the higher wages for other workers. Removing the inferior from work benefited society by protecting American wages and Anglo-Saxon racial integrity.

By pushing the cost of unskilled labor above its value, a minimum wage worked on two eugenic fronts. It deterred immigrants and other inferiors from entering the labor force, and it idled inferior workers already employed. The minimum wage *detected* the inferior employee, whether immigrant, female or disabled, so that he or she could be scientifically dealt with. All civilized societies, Sidney and Beatrice Webb declared, removed their "industrial invalids" from the labor force.[96]

So identified, the inferior workers could be returned to their homes (in the case of mothers not otherwise deficient) or brought under the surveillance of the state—institutionalized, segregated in rural colonies, or even sexually sterilized. Charles R. Henderson, the University of Chicago pastor and sociologist, endorsed banishing the unemployable (under guard) to rural labor colonies, for state direction of their "imperfect labor."[97] Banishment of defectives to "celibate colonies" was the only method of social selection, Henderson wrote, "worthy of the name of rational."[98] As the Webbs had put it, when it came to the unemployable, minimum-wage-induced unemployment was "not a mark of social disease, but actually of social health."[99]

If segregation was insufficient protection, Henderson proposed forcible sterilization. When supporting a large number of the obviously unfit, the commonwealth has the right, Henderson argued plainly, to "deprive them of liberty and so prevent their propagation of defects and thus the perpetuation of their misery in their offspring."[100] Henderson defended involuntary sterilization on grounds that the alternative, demanding more children from the fit, was unfair and unrealistic.

Columbia's Henry Rogers Seager, future AEA president (1922) and a leading progressive economist, is sometimes celebrated as the father of Social Security for his influential Progressive Era research advocating compulsory social insurance. During the minimum wage campaign of the 1910s, Seager joined the battle by arguing that American workers needed protection from the "wearing competition of the casual worker and the drifter" and from the other defectives dragging down their wages.[101]

The minimum wage protected the American worker by making it illegal to hire the unfit, those incapable of earning a living wage. The operation of the minimum wage, Seager explained, "merely extend[ed] the definition of defectives to embrace all individuals, who even after having received special training, remain incapable of adequate self-support."[102] It was healthier for the society, Seager's economics textbook taught, to make the unemployable objects of public charity than to allow them to compete with their betters for jobs.[103]

Seager, like Henderson, made clear that public charity was not a permanent solution for those who, after training, could not earn the legal minimum. "If we are to maintain a race that is to be made of up of [the] capable, efficient and independent," Seager warned, "we must courageously cut off lines of heredity that have been proved to be undesirable by isolation or sterilization."[104] Such eugenic pruning, Seager wrote, ensured the American population would grow from the top and not from the bottom.[105]

A. B. Wolfe, an American progressive economist and also a future AEA president, likewise argued for the eugenic virtues of removing from employment those who were a burden on society. Wolfe's term for the unemployable was "the inefficient." He made clear that he was not disposed to waste much sympathy on the inefficient, saying that their elimination was consistent "with the spirit and trend of modern social economics." The real policy question, said Wolfe, was whether it was better to permit the inefficient to drag down the wages of the "normal worker" or to prohibit their employment, setting them aside as was done with backward and subnormal school children.[106]

The minimum wage barrier not only protected "American" wages by deterring potential competitors, it also identified the unemployable already working. A legal minimum, as the Webbs put it, marked out weaklings and degenerates, "so that they could be isolated and properly treated."[107] Sidney Ball, a fellow Fabian socialist, likewise lauded the minimum wage because it sifted the "industrial residuum," which permitted "restorative, disciplinary, or, it may be, surgical treatment."[108]

Harvard's Arthur Holcombe, a key figure in the progressives' minimum wage campaign of the 1910s, was appointed as an expert to the Massachusetts Minimum Wage Commission, the first such commission in the United States. Holcombe held up the minimum wage's power to identify the unemployable class, facilitating the task of "giving its members treatment suitable to their condition."[109] Social economist Edward Cummings, also at Harvard, applauded the minimum wage for its power to clearly isolate the unemployable, so that they might be scientifically dealt with.[110]

Felix Frankfurter, the AALL's legal counsel after Louis Brandeis was appointed to the US Supreme Court, invoked the segregating effects of minimum wage laws to justify his defense of Oregon's minimum wage law. Frankfurter argued that the states' police power permitted them to override the individual's right to freely contract in the name of protecting society's health, welfare, or morals. Because a successful minimum wage sorted "the normal self-supporting worker from the unemployables" (by idling them), it served a compelling state interest in public health. The minimum wage, Frankfurter suggested, was but a first step toward the solution of determining "how to treat those who cannot carry their own weight."[111]

The economists among labor reformers well understood that a minimum wage, as a wage floor, caused unemployment, while alternative policy options, such as wage subsidies for the working poor, could uplift unskilled workers without throwing the least skilled out of work. Royal Meeker (1873–1953), a Princeton economist trained at Columbia by E.R.A. Seligman, opposed subsidizing poor workers' wages on these very grounds. Eugenically minded progressive economists such as Meeker preferred the minimum wage to wage subsidies not in spite of the unemployment the minimum wage caused but because of it.

Meeker served as Woodrow Wilson's Commissioner of Labor from 1913 to 1920 and founded the International Labor Organization in 1920. One of the world's most senior labor officials supported the minimum wage because he believed it was better for the state to "support the inefficient wholly and prevent the multiplication of the breed than to subsidize incompetence and unthrift, enabling them to bring forth after their kind."[112]

Some prominent Progressive Era eugenicists mistook the minimum wage for a guaranteed income. Their (mistaken) fear was that a minimum wage would pay workers more than they were worth, making it possible for men of inferior heredity to support a family and thus produce more deficient offspring. Paul Popenoe, editor of *The Journal of Heredity*, and Roswell Johnson warned of this dysgenic outcome in in their influential *Applied Eugenics* text.[113]

But a minimum wage did not guarantee a level of income; it only made it illegal to hire workers at a rate below the minimum. A minimum wage thus offered income only to those able to command the minimum, while denying it to those whose labor was worth less.

* * * * *

The affinities between eugenics and labor reform help explain why so many progressives were drawn to eugenics—both movements shared vital commitments to anti-individualism, social control, efficiency, and the authority of scientific experts. The progressives were not alone in their illiberal tendencies; conservatives and socialists also embraced eugenics in the Progressive Era, reminding us of a point made in Chapter 3. One branch of American conservative thought supported vigorous national government and was glad to subordinate individual rights to the state's reading of the social good.

If some Progressive Era conservatives were more skeptical of using state power to regulate human breeding, others were not. Frank Taussig and Frank Fetter, two celebrated founders of American economics, illustrate the point.

Frank Fetter (1863–1949) was a distinguished professor of economics at Cornell and Princeton Universities. At Princeton, where he taught from 1911 to 1933, Fetter was the first chairman of its Department of Social Economics. The AEA made him its ninth president in 1912. Fetter is sometimes regarded as part of the Austrian tradition in economics, which emphasized the subjective nature of value, celebrated the virtues of free markets, and viewed state power with skepticism.

Fetter nonetheless believed that when it came to the threat of the unfit, laissez-faire was untenable. Though free markets brought higher living standards, this very progress suspended the "old brutal elimination of the unfit," permitting "the ignorant, the improvident, the feeble-minded" to contribute more of their defective heredity to future generations. Only the science of eugenics could prevent the imminent arrival of "the noontide of humanity's greatness."[114]

There was no doubt, Fetter's economics textbook asserted, that the US population was increasing more from the "less provident, less enterprising, less intelligent classes," whose multiplication was extinguishing the more capable. The "rational direction ... of perpetuating the race," Fetter wrote, was needed to stem the rising tide of disease, weakness and degeneracy.[115]

Frank Taussig (1859–1940) was professor of Economics at Harvard, where he edited the *Quarterly Journal of Economics* for forty years. An expert in international trade, Taussig was elected president of AEA in 1904, and Woodrow Wilson made him the first chairman of US Permanent Tariff Commission in 1917. Taussig served as Wilson's economic advisor at the Versailles Peace Conference in 1919 and continued as a member of the president's Industrial Conference to the end of Wilson's second term. His *Principles of Economics*, first published in 1911, went into multiple editions over nearly three decades.[116] His friend and successor at Harvard, Joseph Schumpeter, lauded Taussig as "the American Marshall," a reference to the preeminence of Alfred Marshall in English economics.

Taussig was lukewarm toward the minimum wage, saying it was impossible to compel employers to retain or hire workers whose labor services were worth less than the minimum. We don't know how many workers will lose their jobs, said Tausig, but "unemployed there will be."[117]

And yet, Taussig mused in his *Principles* textbook, perhaps some of the unemployment created was a merit and not a defect of the minimum wage. When considering the question of "how to deal with the unemployable," Taussig distinguished two classes of the unemployable, the aged, infirm and disabled, and the "feebleminded ... those saturated with alcohol or tainted with hereditary disease ... [and] the irretrievable criminals and tramps."

The first class might be dealt with charitably, but the second class of unemployable, Taussig proposed, "should simply be stamped out." "We have not reached the stage," Taussig allowed, "where we can proceed to chloroform them once and for all; but at least they can be segregated, shut up in refuges and asylums, and prevented from propagating their kind." Generations of Harvard students, and the many others who used Taussig's economics text, were taught that dealing with the unemployable required stern eugenic remedies.[118]

Progressive Era eugenics appealed also to socialist economists, of whom the Wharton School's Scott Nearing provides an illustration. Nearing was greatly influenced by Simon Nelson Patten while studying graduate economics at the University of Pennsylvania. He earned his PhD in 1909 and taught economics and sociology at the Wharton School for nine years. A radical, Nearing's social activism inflamed the businessmen of the University of Pennsylvania's Board of Trustees, and he was dismissed in 1915. Three years later, Nearing's antiwar activism led to his indictment under the notorious Espionage Act. Nearing is remembered for his long life, his radical politics, and his boldness in opposing conventional wisdom.

In one respect, however, Nearing was utterly conventional—his enthusiasm for eugenics. In his years at Wharton, Nearing published articles on race suicide and on the distribution of "American Genius" for *Popular Science Monthly*. He and Nellie Nearing proselytized for eugenics in the pages of *Ladies Home Journal.*[119] Eugenic themes pervade Nearing's *Social Sanity*, in which he declared unequivocally, "persons with transmissible defects have no right to parenthood and a sane society in its effort to maintain its race standards would absolutely forbid hereditary defectives to procreate their kind."[120]

Nearing found American capitalism intolerable, so he turned his eyes to the better future promised by the social control of human mating. With a nod to both Friedrich Nietzsche and George Bernard Shaw, Nearing summarized his eugenic views in *The Super Race: An American Problem* (1912). In it, Nearing celebrated the ancient Greek practice of eliminating unfitness "by the destruction of defective children." Modern ethics might deplore such a custom, but Americans must recognize its end "as one essential to race progress." Denying the right of parenthood to defectives, Nearing wrote, would more humanely do the necessary work once done by infanticide.[121]

Nearing went on to assert that permitting perpetuation of hereditary defects was "infinitely worse than murder." After all, the murderer "merely eliminates one unit from the social group," whereas the transmission of defective heredity curses and burdens "untold generations." A truly just society thus had an overwhelming obligation to prevent the crime of defective heredity. For the price of six battleships, Nearing estimated, the United States could house, in isolation, all of its defectives. Such a policy would remove, at a stroke, the "scum of society" and, by preventing defectives from procreating, end their threat to future generations.[122]

Conservatives like Fetter and Taussig and radicals like Nearing had little in common ideologically, but they all shared an enthusiasm for eugenic policies. They remind us that Progressive Era eugenics and its discourses of heredity and hierarchy influenced conservative and radical intellectuals as well as the progressives. The notion that inferiors threatened American workers, the Anglo-Saxon race, and American democracy had enough currency to be commonplace in economics textbooks, in which economists of nearly every stripe praised the eugenic virtues of regulating immigration, hours, wages, marriage, contraception, and the right to have children.[123]

In their textbooks as in their other publications, economists, like so many Progressive Era intellectuals, identified inferior heredity with low intelligence, illiteracy, vice (drinking, gambling, and prostitution), pauperism, race, ethnicity or national origins, and labor so unskilled as to be unable to command a minimum wage. Inferiority was also identified with gender.

The progressives' case for restricting women's employment, couched as it often was in a paternalistic language of protection, was subtler than the eugenic hysteria directed at immigrants and defectives. Nonetheless, as with other groups they deemed unemployable, leading progressives portrayed women's employment as destructive—a threat to the wages of male heads of household, a threat to the sanctity of the home, and a threat to the eugenic health of the Anglo-Saxon race.

10

Excluding Women

WOMEN AS WORKERS AND REFORMERS

In 1910, women accounted for 21 percent of the US labor force and 45 percent of professional employment, owing to their predominance in the teaching profession. Millions of American women worked for wages in the Progressive Era. The poor, especially African American women, had no other choice. But the progressives, including the women among them, were profoundly ambivalent about women's employment. Gender, no less than race and intellect, deeply and invidiously informed labor reform.

Between 1909 and 1919, forty American states restricted working hours; fifteen imposed minimum wages, and all but nine paid stipends to single-parent families with dependent children.[1] This outpouring of progressive legislation is rightly regarded as a cornerstone of the American welfare state. Yet maximum hours, minimum wages, and mothers' pension laws applied to women and women only. Male workers were exempted.[2]

Progressive women were at the forefront of American labor reform. In the 1910s, issues of family, child rearing, reproduction, and the maintenance of social morality were widely regarded as the province of women, and progressive women achieved a far-reaching political influence in an era when many could not vote.

In key fields of state regulatory authority, such as factory inspection and child welfare, progressive women were the experts. When, in 1912, Congress established the Children's Bureau in the US Department of Labor, Julia Lathrop was tapped to direct it, the first woman to head a federal agency. Lathrop hired other women to staff her bureau, many of whom, like Lathrop, were experts with long tenures in the immigrant settlement houses.[3]

PROTECTING WOMEN FROM EMPLOYMENT

Progressive women like Lathrop were well represented among those who agitated for gendered legislation, that is, labor laws for women only. Their case for regulating women's employment, couched as it often was in the language of protection, was subtler than the eugenic hysteria directed at immigrants and the feeble-minded. But progressive portraits of women's employment were nonetheless informed by eugenic fears, and they displayed the fundamental contradiction at the core of progressive reform more generally: simultaneously depicting women as helpless victims in need of state uplift and as dangerous threats requiring state restraint.

The progressive case for protecting women began with female difference, and difference usually meant inferiority. The claim was that women, as the biologically weaker sex, needed (like children) special protection from the demands of employment, usually in the form of restrictions on hours or bans on night work.[4] Richard T. Ely, like many progressives, argued that night work should be forbidden for women, as should any type of employment "injurious to the female organism."[5]

The "weaker sex" discourse was hardly new to America, but progressives, famously in the Brandeis brief, put it to new use justifying state regulation of women's employment. Compiled by Josephine Goldmark and Louis Brandeis for the US Supreme Court case *Muller v. Oregon* (1908), the Brandeis brief was commissioned by Florence Kelley of the National Consumers League and funded by the Russell Sage Foundation.

The brief supported Brandeis's defense of an Oregon statute that restricted women's hours to no more than ten per day and sixty per week. The Brandeis brief is famous, because it appealed to evidence from social science, not just legal precedent. The idea of appealing to scientific evidence was novel, but the brief itself was just a hodgepodge of expurgated articles, culled from a rummage through a mass of labor investigations, with conclusions friendly to the client.

Goldmark, Brandeis's sister-in-law, began by asserting "the special susceptibility to fatigue and disease which distinguishes the female sex qua female."[6] She claimed that women were fundamentally weaker than men, in terms of strength, energy, attention, and application. One expert Goldmark cited, English sexologist Havelock Ellis, declared that women's physical inferiority was explained by the fact that there was more water in women's blood than in men's.[7]

Goldmark's case for women's special susceptibility combined an appeal to science, however dubious its sources, with conclusions supporting traditional gender prejudices. Its purpose was to persuade the *Muller* court that long hours were "more disastrous to the health of women than to men," so it omitted all evidence that men might also suffer from fatigue and overwork. Brandeis's appeal to the objectivity of science was not itself objective; he was making a case.

Oregon won its case, and the *Muller* decision was a landmark victory for progressive labor legislation. The Supreme Court affirmed the constitutionality of labor laws for women. But Kelley, Brandeis, and Goldmark were playing a dangerous game.

If their arguments for women's inferiority succeeded, they risked inscribing into law the subordinate status of women in the economy and in the polity. If their argument that courts should entertain scientific evidence succeeded, they risked the likelihood that less selective evidence would point to sex equality rather than sex difference. Thus, if the claim of women's inferiority withstood judicial scrutiny, women's equality was threatened, and if the claim did not withstand scientific scrutiny, labor legislation for women only was threatened.

A second influential progressive justification for protecting wage-earning women arose during the US minimum wage campaign of the 1910s: the protection of women's sexual virtue. Prostitution, known euphemistically as "the social vice," was a singular preoccupation of progressive reformers, especially those with ties to the social purity movement. Columbia economist Henry R. Seager called "prostitution in aid of inadequate wages" the "greatest disgrace of our civilization."[8]

On its face, the idea was plausible; better-paid workers were less likely to succumb to using prostitution to bolster their incomes. Seager claimed that the "$8-a-week girl had more power to resist the temptations which our cities constantly present than the $5-a-week-girl," a sentiment widely held and found in later versions of the Brandeis brief. But what of the workers who lost their jobs because of a higher minimum?

John Bates Clark scolded his junior colleague, reminding Seager that if $5 a week forced a person into vice, then "no wages at all would do it more surely and quickly."[9] Clark was angry, because Seager was well aware of this argument. Seager himself had praised the minimum wage as a tool for removing mental defectives and other unemployables from employment, but he offered no provision for the women who would be idled. Seager simply presumed that women would be economically supported and their virtue protected by husbands, fathers, and brothers.

PROTECTING EMPLOYMENT FROM WOMEN

When proposing to protect women, progressives portrayed women as weak and defenseless. But when proposing to control women, they portrayed women as dangerous threats to their husbands, children, and the health of the race. Indeed, progressive arguments for regulating women's employment invoked women's obligation to society at least as often as it did society's obligation to women. As Chicago economist and reformer Sophonisba Breckenridge (1866–1948) recognized in 1906, regulation of women's employment was not "enacted exclusively, or even primarily for the benefit of women themselves."[10]

The tension between protecting women from employment and protecting employment from women manifested itself regularly. Leading the campaign for minimum wages for women, Florence Kelley cited the success of the Victoria, Australia, minimum wage law. The Victoria minimum wage law, enacted in 1894, was one of the first of its kind. Labor reformers worldwide made the long journey to the antipodes to study its workings. The Australians succeeded, said Kelley, because the minimum wage eliminated the "unbridled competition" of women, children, and Chinese, who had been "reducing all the employees to starvation."[11]

Kelley's formulation excluded women workers from the category of "employee," grouping them with children as exploited victims in need of state protection. Women and children were, she said, "the weakest and most defenseless breadwinners in the state."[12] At the same time, Kelley also placed women with the Chinese, whom she represented as low-wage threats to white Australian men. Thus, Kelley accused women of undercutting white men. The low female standard, like the low Chinese standard, made women workers a competitive danger requiring state restraint. The tension between protecting women and restraining women was often overlooked, in part because the remedy was same in either case: removal from employment.

The claim that women's employment was dangerous to others generally took one of two complementary forms. The first class of argument, "the family wage," still resonates today. The family-wage argument said that women's employment was bad for their families. It undercut the wages of their bread-winning husbands (indeed, all men), and it also threatened their children's wellbeing. The second class of argument, the "mothers of the race" argument, reflected the deeply engrained hereditarian thinking of the day. It said, as mothers of the race, women were obliged conserve human heredity and should not risk the race's health with overwork and fatigue (at least not with *paid* overwork and fatigue).

The family-wage principle portrayed employed women as usurpers of jobs that rightfully belonged to men. As Florence Kelley put it, any industry that hired women instead of men was socially subnormal.[13] Returning women to the home had the additional benefit of ensuring that women properly carried out their eugenic duties as mothers of the race.

Premised on traditional sex roles, the family-wage and mothers-of-the-race principles argued not for women's rights but for their obligations, not for women's welfare, but for the welfare of men, children, and the race. Nonetheless, the family-wage and mothers-of-the-race principles proved immensely popular among progressive labor reformers, not least the women among them.

THE FAMILY WAGE

The family-wage argument, unlike the vitriol directed at immigrants and defectives, valorized women. As the US Children's Bureau put it, "the welfare of the home and family is a woman-sized job in itself."[14] But "maternalist" labor reformers celebrated women's work only insofar as it was confined to the maternal sphere. Men were providers, heads of household entitled to wages sufficient to support a family, and women were mothers whose place was in the home.

Maternalists well understood that poor women desperately needed what gainful employment they had, but they also maintained, with no less vigor, that motherhood and employment were incompatible. Florence Kelley put it plainly: "Family life in the home is sapped in its foundations when mothers of young children work for wages."[15] Kelley lamented a world in which men were no longer breadwinners, and she wagged her finger at immigrant men, whose wives were more likely to be employed, declaring, "the American tradition is that men support their families, their wives throughout life, and the children at least until the fourteenth birthday."[16]

Julia Lathrop, addressing the National Conference of Social Work in 1919, made it clear that women had to choose between motherhood and employment. "Let us not deceive ourselves," the director of the US Children's Bureau said. "A decent family living standard … means a living wage and wholesome working life for the man, a good and skillful mother at home to keep the house and comfort all within it."[17]

Torn between the desire to uplift poor women and the felt need to control poor mothers' choices with respect to family life, maternalist activists devised a three-pronged regulatory plan. First, restriction of women's hours would return mothers to the home. Second, the reduction in female labor competition caused by the first tactic would increase men's wages sufficiently to support a family. Third, women without a male provider would receive mothers' pensions in the form of state payments to fatherless families with children.

Family wagers presumed that women, like immigrants and defectives, had lower standards than did men. Progressive texts were blandly explicit about "American," "immigrant" and "feminine" standards of living.[18] Labor reformers sometimes slipped from the idea that women's lower standards were innate to the idea that women's lower standards were culturally caused. But such slippage was rarely noted, and with the assumption of Lamarckian inheritance, mattered little for the theory's causal story.

Henry R. Seager justified his theory of the lower female standard by appealing to the idea of women's "natural" dependency on a male provider. Women, Seager claimed, accepted low wages because they were working only to lighten the burden of their dependency on their husbands, fathers, and brothers.[19]

Ely went so far as to deny that women's earnings positively contributed to their families' incomes. "A man and a wife working together secure no greater wages than the man alone in industries in which women are not employed," Ely asserted, assuming that the economic competition of women drove men's wages down so much as to completely offset the additional family income women provided.[20] Ely offered no evidence for his claim.

John R. Commons's study of the economic effects of immigration, conducted in 1900 as a special agent for the US Industrial Commission, condemned immigrants for undercutting American workers, for impeding union organization, and, of course, for outbreeding their Anglo-Saxon betters. Commons also found space to accuse women of undercutting and displacing breadwinning American men. Women, Commons asserted, were unemployable on account of their "carelessness, ill temper and unreliability."[21]

The contradiction of labor reformers' portraying wage-earning women as both exploited victims and dangerous threats made for an unstable rhetorical amalgam. Labor economist Helen Sumner's 1910 report to Congress, "Condition of Woman and Child Wage-Earners in the United States," illustrates well the progressives' ambivalence toward women's employment. Sumner (1876–1933) had been trained by Commons at the University of Wisconsin, and she made significant contributions to the monumental *History of Labour in the United States* (1918). Her *Labor Problems* textbook sold briskly, and she later became Julia Lathrop's second in command at the US Children's Bureau.

Sumner's report to Congress began sympathetically. The story of women in industry was one of long hours, low wages, unsanitary conditions, overwork, and monotony. Wage-earning women deserved sympathy and state uplift. Just a few sentences on, Sumner struck a very different tone. The story of women in industry, she wrote, was one of underbidding, of strike breaking, and of lowering "standards for men breadwinners." The very same women also deserved contempt and state restraint.[22]

Father John A. Ryan, a leading progressive in labor reform circles, argued for a natural right to a living wage. Pope Leo XIII had opened his church's doctrinal door to economic reform when he published *Rerum Novarum: The Condition of Labor*, and Ryan seized the chance to promote a reform Catholicism based on the God-given rights of workers. But not all workers.

Ely published Ryan's *A Living Wage* in his influential *Citizen's Library* series, and Ryan became one of America's most prominent spokespersons for a living wage. He later drafted Minnesota's minimum wage statute and made a name as leading progressive intellectual.[23]

Ryan defined a living wage differently for men and women. Since nature decreed that men were heads of their households, men were entitled to a family wage.[24] Women could not be heads of household, though many in fact were, so a women's living wage was not a family wage. A employed women deserved only enough to support herself.[25]

Whatever nature decreed, Ryan continued, women's physical wants were simpler. Centuries of drudgery and suffering had taught women to survive on less, which explained why women were willing to accept low wages. Women's lower standards thus threatened men, who had families to support.

Ryan's logic of wage undercutting led him to spot an incentive problem with his scheme. If family men were entitled to a family wage, then firms would prefer to hire single men, who could be paid less. The result would be adverse selection for single men, a "very undesirable kind of celibacy." Ryan's solution was to award every man a family wage, whether or not he had a family.[26]

Ryan's answer to the economic competition of single men was to award them higher pay. His answer to the economic competition of women was to remove them from the labor force. Ryan put it plainly: the welfare of the family and of society "renders it imperative that the wife and mother should not engage any labor except that of the household"[27]

Ryan's reform Catholicism, with its emphasis on natural rights, was exceptional among labor reformers, but his claim that employed women wrongly undercut male workers was a commonplace. Rheta Childe Dorr, a muckraking journalist who went undercover to investigate the lives of women in industry, was another labor reformer who—notwithstanding her obvious sympathy for the working-class women toiling in factories and shops—portrayed them as low-standard competitors to men. Even as she advocated for shorter hours and better workplace conditions for women, she decried "the woman's invasion" of industry, enabled by her "irresponsible cheapness."[28]

The wage-earning woman, she wrote, was "the white Chinaman of the industrial world. She wears a coiled-up queue, and wherever she goes, she cheapens the worth of human labor."[29] This ugly metaphor perfectly captured the capaciousness of the "unemployable" discourse, deploying racist analogy to discredit millions of wage-earning women as coolie invaders, whose whiteness could neither disguise nor redeem the threat their inferiority posed to the American workingman.

Not all progressives bowed to the family-wage construct. Political journalist Walter Lippmann, writing in the *New Republic*, assailed the idea that women's wages were low because women had low standards. Women weren't cheap, Lippmann thundered, they were exploited. Women's wages were determined by the shop's custom, the whim of the boss, and by arbitrary decision. Sweatshop operators, Lippmann wrote, were too incompetent to know how much an individual woman was producing, so it could only be by chance she was getting paid the value of her addition to output.[30]

Among labor reformers, two influential progressives rejected family-wage doctrine: Mary Van Kleeck (1883–1972) and Sophonisba Breckinridge. Van

Kleeck, then director of Industry Studies at the Russell Sage Foundation, argued that women needed special legal protection not because of their biological or intellectual limitations, but because they were grossly and unfairly underpaid.

Low wages were destructive and exploitive whether the worker was male or female, but wage-earning women were even more exploited than were impoverished male workers.[31] Women were uniquely vulnerable, because, barred from most labor unions, they lacked bargaining power with their employers. It was rock-bottom wages rather than innate female difference that made women's employment a special concern of the state.

Van Kleeck's sensible argument did not prevent others, such as Edward A. Ross, from ascribing women's lack of bargaining power to innate female difference. Ross said that women lacked bargaining power with employers, because "they are *women.*" A woman's nature made her less aggressive and less willing to strike back at employers by organizing into unions. What was more, Ross said, women were myopic. They always had one eye on marriage, and thus (Ross assumed), exit from employment. Why bother with organizing and union dues when a husband was soon to ride to the rescue?[32]

Sophonisba Breckinridge, like Van Kleeck, came to see the shortcomings of the family-wage principle. A protégé of Ernst Freund, Breckinridge was the first woman to earn a PhD in Political Science from the University of Chicago, where she was also the first woman to obtain a law degree. A pioneering social scientist, Breckinridge taught social economics at the University of Chicago and was instrumental in founding and building the Chicago School of Civics and Philanthropy, which was ultimately merged into the University of Chicago to form its School of Social Service Administration.[33]

Breckinridge once wrote in the maternalist vein of her Hull House colleagues, but came to see how the family-wage principle disadvantaged wage-earning women.[34] If, unlike men, women were not regarded as heads of household, the living wage they were entitled to was only enough to support a young woman living alone. Though many employed women were, in fact, supporting dependents, state minimum wage commissions, abiding by the family-wage construct, set minimums for women at levels inadequate for them to support more than themselves.[35]

Van Kleeck and Breckinridge rightly saw how family-wage doctrine disadvantaged women, but their argument fell on deaf ears. Very few progressives defended labor legislation for women without appealing to the family-wage rationale that women support their families as mothers and not as breadwinners. As historian Linda Gordon concluded, "almost all welfare activists, male and female, endorsed the family-wage principle and considered that women's employment was a misfortune or a temporary occupation before marriage."[36]

MOTHERS OF THE RACE

The progressives' lawyers liked the mothers-of-the-race argument much better than the women-are-even-more-exploited argument. If exploitation was the problem, then special legal protection for women was harder to justify. Exploitation was a matter of degree. But only women could be mothers, and the state clearly had an interest in the health of the race, so motherhood could justify a compelling state interest in regulating women's employment.

The premises of the mothers-of-the-race argument were paternalistic and neo-Lamarckian: wage-earning women could not be trusted with the health of their own children, and overwork adversely affected not only a woman's health but also her heredity. In fact, when push came to shove, labor reformers worried more about hereditary health than women's health. The Brandeis brief made it plain. The most compelling justification for restricting women's working hours was not to improve industrial efficiency nor even to protect women's health or virtue. The most important justification for restricting women's hours was to protect "the woman's fitness for motherhood."[37]

Irving Fisher made the claim in his eugenicist *National Vitality: Its Waste and Conservation.* There were many good reasons for a shorter workday. It protected women's health and there was even some evidence that workers increased their labor productivity enough to compensate their employers for the shorter day. But the most important justification, Fisher declared, was to be found "the interest of the race."[38]

John R. Commons drafted Wisconsin's 1913 minimum wage statute and served on its Minimum Wage Board for more than twenty-five years. The minimum wage for women, he wrote, protected the "welfare of the race and the nation."[39]

Progressive Era women defied traditional gender roles by going to work and to college in increasingly large numbers. Eugenicists bemoaned both trends and found eugenic grounds for condemning working-class factory girls and privileged college-educated women alike. The young woman employed in industry was accused of "unfitting" herself for motherhood, risking degenerate progeny. The college-educated woman, who delayed marriage and children, was accused of abetting race suicide. The "new woman," whether a factory hand or a privileged alumna, threatened American racial health.

In *Muller v. Oregon* (1908), the US Supreme Court upheld the constitutionality of restricting women's hours in part by invoking Brandeis's mothers-of-the-race argument. The state, Justice Brewer wrote for the majority, could justify its special interest in women's employment on grounds of preserving "the strength and vigor of the race." The hours laws infringing on a woman's right to contract, Brewer said, were "not imposed solely for her benefit, but also largely for the benefit of all."[40]

Shortly after the *Muller* decision, Edward A. Ross took to the pages of *The New York Times* to encourage more restrictive legislation for women. Ross estimated that nearly one-third of unmarried women aged fifteen to twenty-five were employed. Like the maternalists, Ross warned that factory girls, not understanding the health risks of employment, were unfitting themselves for motherhood.

Employing young women, Ross wrote, masculinized them. Absent regulation, the "high-strung, high-bred, feminine type which is our pride" will be displaced by the "Flemish-mare type" of woman, the only female type able to withstand the rigors of the factory. In but three or four generations, Ross told *Times* readers, employment will make the lower stratum of American women "squat, splay-footed, wide-backed, flat-breasted, broad-faced, [and] short-necked."[41]

Progressive feminists who advocated the overthrow of traditional sex roles clearly rejected gender constructs such as Ross's. But they too appealed to eugenic concerns, as suggested by the example of Charlotte Perkins Gilman and her sui generis feminist eugenics.

Gilman blamed androcentric culture for the subjugation of women, which had injured women and thereby the Anglo-Saxon race. Generations of male domination had increased the proportion of weak, unfit women, which, in turn, caused race degeneration. "A race of women who are contented to be cooks and housemaids," Gilman wrote, "do not give as noble a motherhood as the world needs."[42] Liberating women from male oppression, Gilman contended, would make women stronger and fitter and would thereby improve the race.

At the same time, Gilman urged women to assume responsibility for their "measureless racial importance as makers of men." Gilman's told women that their role was "to improve the race by right marriage." The state, for its part, would help women select fitter mates, by certifying men's biological fitness.[43]

Gilman was more feminist than maternalist. She wanted to emancipate women from domestic drudgery and the burden of bearing and rearing many children, not return them to it. And yet, as radically as she reconceived family life, Gilman still represented motherhood in terms of race progress.[44]

Working with very different premises, Theodore Roosevelt also represented motherhood in terms of race progress. Gilman had identified oppression of women as a cause of race degradation and liberation of women as the remedy. Roosevelt identified women's selfish shirking of reproductive duty as the cause of race suicide and more children as the remedy. Motherhood, Roosevelt declared, was a woman's "primal and most essential duty."[45]

The women Roosevelt accused of being "race criminals" came from privileged backgrounds. Their families did not have lower fertility forced on them by the economic competition of working-class immigrants or defectives. The college-educated woman had a choice whether to have fewer children, and when she chose to have fewer, Roosevelt condemned her as decadent and selfish.[46]

Gilman and Roosevelt were antagonists with opposed positions. Gilman blamed race degradation on the subjugation of women, while Roosevelt accused women of abetting race suicide. Gilman wanted to liberate women of good stock from the demands of maternity, and Roosevelt wanted to conscript them into it. And yet both Gilman and Roosevelt represented motherhood in terms of race progress, warning of eugenic danger in support of their antagonistic causes.

Economist Albert B. Wolfe spent two years living in Boston's South End immigrant settlement, a sojourn funded by a Harvard fellowship. Wolfe warned in 1906 of the dysgenic consequences of lodging houses—lodging establishments for respectable, unmarried middle-class men and women. Lodging-house clients, Wolfe asserted, were superior to the tenement dweller in physical vitality, education, and ambition, but being "anchored indefinitely" in the lodging house, they were producing far fewer children than was the immigrant. The sooner marriage could rescue the young men and women from the lodging house's "sophisticating, leveling and contaminating influences" the better, Wolfe asserted. America needed more children from this higher class to offset the immigrant-caused decline in American physical, mental, and moral vitality.[47]

Simon Nelson Patten issued a contemporaneous race-suicide warning in "The Crisis in American Home Life." The great Philadelphia middle class, Patten warned in 1910, had become "homeless," meaning they lived in lodging houses. Patten warned that if these middle-class families remained childless or produced only one child, the certain outcome was that Philadelphia would be overrun by its recent immigrants. The Quaker, the Puritan, and the Scotch-Irish, Patten lamented, would soon cease to exist.[48]

The number of women enrolled in America colleges tripled between 1890 and 1910. College-educated women married less often, married later, and had fewer children than did less privileged women. In 1917, of all the graduates of women's colleges, only half were married.[49]

A scholarly cottage industry sprang up to investigate and deplore the declining fertility rates of the American elite.[50] Nellie Nearing, for example, found that women's colleges produced less than one child per alumna. Irving Fisher's address to the Eugenics Research Association warned, "the average Harvard graduate is the father of three fourths of a son and the average Vassar graduate the mother of one half of a daughter."[51]

Fisher identified contraception as the cause of race suicide among the educated and well-to-do classes. As practiced, Fisher said, birth control was dysgenic; it was used least where it was needed most. Only when contraception was successfully extended to the lower classes would birth control become a eugenic tool.[52] Until such time, progressives added "willful sterility" to their already long list of eugenic anxieties.

Male alumni were also having fewer children. Yet women were singled out for neglecting their racial duties, because women graduates exhibited even lower birth rates[53] and because women were held responsible for improving the race. Scott and Nellie Nearing claimed that eugenics had shown conclusively that the future of the race rested on women's shoulders. Women must select, and on their selection the progress of the race depended. A woman's "whole soul, conscious and unconscious," the Nearings wrote, was "best conceived as a magnificent organ of heredity."[54]

* * * * *

American labor reformers found eugenic dangers nearly everywhere women worked, from urban piers to home kitchens, from the tenement block to the respectable lodging house, and from factory floors to leafy college campuses. The privileged alumna, the middle-class boarder, and the factory girl were all accused of threatening American racial health.

Paternalists pointed to women's health. Social purity moralists worried about women's sexual virtue. Family-wage proponents wanted to protect men from the economic competition of women. Maternalists warned that employment was incompatible with motherhood. Eugenicists feared for the health of the race.

Had she wanted to abide the various injunctions of labor reformers, the employed woman could not have done so. If she were paid very little, she was admonished for endangering her health, risking her virtue, and threatening hereditary vigor. If she commanded a slightly higher but still modest wage, she was condemned for undercutting men's family wages and for neglecting her maternal duties. If she were well paid, she was admonished for selfishly acquiring an education, pursuing a career, and thus shirking her reproductive responsibilities to society and the race.

Motley and contradictory as they were, all these progressive justifications for regulating the employment of women shared two things in common. They were directed at women only. And they were designed to remove at least some women from employment.

WHY WOMEN ONLY?

Historians debate what drove Progressive Era labor reformers to become standard-bearers of a woman-only regime of labor regulation. Why did so many progressives, especially the women among them, reinforce rather than confront traditional notions of women as the inferior sex, weak, defenseless, selfless, and meant to be at home?

One explanation is that maternalism was fundamental to the outlook of the progressive women who had "mothered" entire communities in their settlement work. Maternalism was just too integral to their outlook to be shed altogether. Another explanation argues that labor reformers adopted maternalist rhetoric only to placate a reactionary US Supreme Court, which, in *Lochner v. New York* (1905), signaled its unwillingness to uphold state regulation of male labor contracts. On this reading, all the talk of female inferiority and the incompatibility of motherhood and employment was a ruse, a tactic made necessary by American constitutional politics.[55]

A third explanation says that maternalism was a response to the occupational barriers faced by highly educated and ambitious women, who were given little access to male-dominated universities and related professions. The price of carving out a professional territory of their own, and of building a "feminine" expertise among poor women and children, was to abide and endorse traditional ideals of female behavior—service, self-sacrifice, and devotion to children and family.[56]

Whatever mixture of motives lay behind the progressive arguments for a gendered regime of labor legislation, the consequences were clear.[57] Whether labor reformers were maternalists, or retailed maternalism to accommodate what were very real constitutional and vocational constraints (or both), the legal strategy of female inferiority ran aground, calamitously, with the *Adkins* decision in 1923.

Having upheld fifteen years of women's labor legislation since the pivotal *Muller* decision, the US Supreme Court reversed field, striking down a District of Columbia minimum wage law on the grounds that women's newly won suffrage rights elevated their legal status from protected inferiors to freely contracting citizens. Women had the vote and could fend for themselves, said the *Adkins* Court.

Somewhat less noticed was the majority's subtle shift to a different economic theory of wage determination. Implicitly, the Court rejected the progressive idea that wages were determined by living standard and instead adopted a neoclassical theory of wages. Justice Sutherland allowed that state laws could require employers to pay workers "the value of services rendered," but a law that insisted that wages be determined by a worker's necessities referred to "circumstances apart from the contract of employment" and thus was unconstitutional.

Labor, Sutherland wrote, was a commodity like any other:

> if one goes to the butcher, the baker or grocer to buy food, he is morally entitled to obtain the worth of his money, but he is not entitled to more. If what he gets is worth what he pays, he is not justified in demanding more simply because he needs more, and the shopkeeper, having dealt fairly and honestly in that transaction, is not concerned in any peculiar sense with the question of his customer's necessities.[58]

In dissent, Chief Justice William Howard Taft protested that the Court lacked standing to pronounce on questions of economics. What was more, Taft said, woman suffrage did not somehow increase women's physical strength, nor did it remove the other female "limitations" recognized in *Muller*. Oliver

Wendell Holmes, Jr., agreed. "It will need more than the Nineteenth Amendment," Holmes wrote, "to convince me that there are no differences between men and women."[59]

Taft and Holmes, however, were in the minority, and *Adkins* devastated the progressive legal strategy of founding labor legislation on the inferiority of women. Louis Brandeis, now Justice Brandeis, was forced to recuse himself and could only watch as the *Adkins* version of the Brandeis brief failed for the first time. Speaking for many, an angry and despairing Henry R. Seager denounced *Adkins* as "the most severe blow which progressive American Labor Legislation has yet received at the hands of the Supreme Court."[60]

The fallout from *Adkins* split apart the progressive coalition. Some, such as Felix Frankfurter, held out hope that the *Adkins* majority had left enough daylight for a suitably modified strategy of female inferiority to pass constitutional muster once again. Others, including Kelley herself, believed that *Adkins* had ended any prospect for using sex differences to justify US labor legislation for women. She and her like-minded allies proposed amending the Constitution to go around rather than through the recalcitrant Court.[61]

Frankfurter thought an amendment was hopeless, even though progressives had successfully amended the Constitution four times in seven years. Henry R. Seager drew up a constitutional amendment that would require Supreme Court supermajorities (two-thirds or even three-fourths) to overturn labor legislation, presaging Franklin Roosevelt's court-packing scheme in the months prior to the passage of the Fair Labor Standards Act of 1938, which created a federal minimum wage for all workers, male and female.

Even more divisive was the post-*Adkins* defection of equal-rights feminists from the women-only camp. *Adkins* convinced equal-rights feminists, such as Alice Paul and Maud Younger of the National Women's Party, that the women's-inferiority strategy, which they had always seen as wrong, was now impotent as well. The National Women's Party demanded full legal equality for women, not special protection. In the view of Alice Paul, American women laid claims on the state not by virtue of their inferiority to men but by virtue of their equality to them.

The equal-rights feminists were not the only women's constituency with a grievance against protective legislation. The Women's Equal Opportunity League, founded after New York State's prohibition of night work idled thousands of women, also advocated for equality rather than special protection.[62] But the decisive moment arrived with the first Equal Rights Amendment, put forward by Alice Paul, Maud Younger, and the other equal-rights members of the National Women's Party.

The National Women's Party now regarded women-only labor legislation as another form of sex discrimination, ultimately inimical to women's interests. When it was premised on female inferiority, half a loaf was worse than none. The progressive economists and their labor reform allies strenuously disagreed. Full legal equality for women, they charged, would destroy the hard-won legacy of women-only labor legislation.

The progressives vigorously opposed the Equal Rights Amendment. The constitutional reform needed was not equal rights for women, they said. The constitutional reform needed was to rein in the troglodytes on the Supreme Court.

Unable to bridge this irreconcilable difference, Florence Kelly and other labor legislation activists abandoned the equal-rights feminists as enemies. When John R. Commons became President of the National Consumers League in the decisive year of 1923, his top priority was to oppose the Equal Rights Amendment. With few exceptions, progressives in labor reform chose protection over equality. They relaunched the campaign to protect women from employment and to protect society and the race from the employment of women.

EPILOGUE

Before the First World War, the progressive economists' outsized confidence in their own wisdom and objectivity was matched only by their belief in the transformative promise of the administrative state.[1] Their extravagant faith in expertise and scientific government was sustained by a potent and quintessentially American combination of overconfidence and naiveté, a combination that, like so much else, was dealt a blow by the First World War.

When Irving Fisher gave his presidential address to the AEA meetings in December 1918, Armistice had only just ended hostilities in Europe. Fisher's talk, "Economists in the Public Service," was obliged to face the embarrassment that, for more than thirty years, progressives had celebrated Germany as the wellspring of American progressive economics and the inspiration for the American administrative state.

Fisher acknowledged that America owed to Germany the progressive idea of "making economics of service to 'the state.'" But the war's revelations, Fisher continued, "have made us realize, to our horror, that 'the state' served by the German economists ... was simply the Hohenzollern dynasty." The German state, it turned out, was not Lester Frank Ward's ideal of an enlightened people's collective mind. It was, rather, a criminal regime, and the German economists who served it, Fisher said darkly, were prostitutes.[2]

John Stuart Mill had long before warned that politicians will not accept instruction from their technocratic betters, and that even instructed governments can do more harm than good. Fisher and his fellow progressives had ignored Mill's cautions, and now their paragon of good government had pursued evil ends, and its experts had been willing accomplices.

Some progressives became disenchanted with the idea of scientific expertise in the service of an administrative state. Herbert Croly, once the leading publicist of American social control, soured on expert social science and its influence on government policy. Alternately rueful and sardonic, Croly lamented that progressives had too confidently placed their faith in administrative means to democratic ends, believing that society's "better future would derive from the beneficent activities of expert social engineers who would bring to the service of social ideals all the technical resources which research could discover and ingenuity could devise."[3]

In practice, however, social control, the origin of which Croly credited to Lester Frank Ward's *Dynamic Sociology* (1883), lacked the "skeptical modesty of science." The economic experts, whom Croly had once lionized, did not know enough, and they should not have pretended to scientific knowledge when justifying their claim to political power and its grave responsibilities.[4]

Social control, Croly continued, not only claimed to be more scientific than it was. It was also undemocratic. Because economic reform was not of or by the people, it could only be for the people. In the absence of intelligent and active political participation, the people's consent, Croly said, was "fictitious." The social engineer had devolved into a "traditional law-giver who knew what was possible and good for other people and who proposed to mold them according to his ideas."[5]

Croly's essay was not exactly a *mea culpa*, but he could not have missed the irony. The confident prophet of Progressivism, whose *The Promise of American Life* (1909) enlisted so many in the crusade for social control, was now, fifteen years on, enumerating its failures. Croly's manifesto had ignored the danger that the experts might be overconfident, elitist, and without possession of the scientific knowledge they claimed to have. In so doing, his prophecy proved as fallible as those who had tried to carry it out. Embittered, Croly repudiated the past.

Irving Fisher, in contrast, having acknowledged the dangers he too once ignored, effectively dismissed them. Performing some rhetorical jujitsu, Fisher told the AEA: "The very fact that Germany once inspired us toward an economics in the service of the state should spur us now to avoid the nationalistic perversions of that idea which befell our German colleagues." Fisher, having granted that social control hadn't worked out so well in Germany, proceeded to outline a program of greater government ownership and control of the American economy. The American experts, now cognizant of the dangers, could continue as before.

AMBIGUOUS LEGACY

Fisher was mistaken. American economic reform in the Progressive Era had long since fallen prey, again and again, to nationalistic perversion. Progressive labor reformers advocated for Anglo-Saxon men while condemning immigrants, women, and the disabled as low-standard threats to American wages and Anglo-Saxon race integrity. When progressive political reform reduced political participation, the more elitist progressives applauded the exclusion of the races deemed incapable of self-government. The progressives offered uplift to some but exclusion for others, and did both in the name of progress.

As we have seen, many of the progressive leaders who dedicated themselves to social and economic betterment made invidious distinctions among the poor, valorizing some as victims deserving help, while vilifying others as threats requiring restraint. Progressivism's legacy is this strange and unstable compound of compassion and contempt.

Some scholars have treated the progressives' ambiguous legacy by wishing it away. Those who admired the progressives ignored or trivialized the reprehensible and wrote lives of the saints. Those who disliked the progressives ignored or trivialized the admirable and wrote lives of the proto-fascists. But Progressivism is too important to be left to hagiography and obloquy.

Progressivism's braiding together of the admirable and the reprehensible, starts with its veneration of science. There can no doubt that the progressives in economic reform drew deeply on race science and eugenics to distinguish the victims who deserved uplift from the threats who required restraint.

But here we must be careful to avoid the condescension of posterity. Historians of science remind us that the history of bad ideas is as interesting, and as important, as the history of good ones. This is true because any bad idea of historical importance is, almost by definition, an idea that many people thought to be a good idea at the time. Histories of bad ideas show us something about how science works and what happens when it is harnessed to political and economic purposes.

Eugenics and race science are historically important, and during the Gilded Age and Progressive Era many people—most conspicuously the progressives—thought they were good ideas. The events of the intervening century, some of them horrific, have changed our view. Eugenics and race science are now bad ideas, indeed Bad Ideas, which is why twenty-first-century geneticists, economists, sociologists, demographers, physicians, and public health officials remain reluctant to look too closely at their respective disciplines' formative-years enthusiasm for now discredited notions. The very word "eugenics" remains radioactive, and the temptation to dismiss eugenics and race science as inconsequential pseudosciences is ever present.

But eugenics and race science were not pseudosciences in the Gilded Age and Progressive Era. They were sciences, and Progressivism was, first and foremost, an attitude about the proper relationship of science (personified in the scientific expert) to the state, and of the state to the economy and polity.

Eugenics was ubiquitous during the first three decades of the twentieth century. Hundreds and probably thousands of scholars and scientists, including the world's most eminent geneticists, proudly claimed to be eugenicists. They raised millions to fund eugenic laboratories and founded a passel of scholarly journals dedicated to the study of eugenics and eugenic policy. They convinced governments to regulate marriage, reproduction, immigration, and labor in the name of eugenics. Their leaders even believed that eugenics would one day be a kind of scientific religion.

The reading of Progressive Era eugenics as an inconsequential pseudoscience is understandable, given the crimes committed in its name. But this retrospective designation obscures the influence of eugenics, particularly upon the progressives. Progressive Era eugenics was anti-individualistic; it promised efficiency; it required expertise, and it was scientific.

American Progressive Era eugenics was anti-individualist and illiberal. Its raison d'être was the belief that racial health was too important to be left unregulated. The individual's liberty to make her reproductive, marital, labor, and locational choices free from state interference ended precisely at the point where her choices were seen to endanger the health of the race. As Sidney Webb put it, "no consistent eugenicist can be a 'Laisser Faire' individualist unless he throws up the game in despair." The eugenicist "must interfere, interfere, interfere!"[6]

Eugenics was premised also on efficiency, another progressive lodestar. Eugenicists substituted social selection for natural selection based on the belief that natural selection let alone was too slow, too inefficient, too inhumane, and too indifferent to progress. As Francis Galton encapsulated it, "what nature does blindly, slowly and ruthlessly, man may do providently, quickly and kindly."[7]

Eugenics also demanded expertise. The expert's objective determination of the social good replaced a subjective, individual determination of it.[8] The experts not only knew better, they could be trusted to pursue the common good of hereditary health. No conscientious person, Edward A. Ross said, "should have the hardihood to ignore eugenics."[9] And eugenics was scientific. Even Jane Addams, the historiographic personification of progressive commitment to the socially marginal, made eugenic noises, one measure of the popularity and scientific respectability of eugenic ideas during the Progressive Era.[10] Eugenics' combination of anti-individualism, efficiency, expertise, and science did not, of course, ensnare every progressive. But those who could resist were thin on the ground.

The influence of the new sciences of heredity, including eugenics, helped economic reformers distinguish workers who deserved uplift from workers who deserved restraint. But it cannot explain the felt need to make invidious distinctions in the first place.

Here we need to recall the progressives' desire to create an American nationality, to make the United States singular rather than plural. For progressives, we have seen, the United States was not just a collection of states, still less was it a contractual creation of free individuals who called it into being and could dissolve it as well. For progressives, the United States was an organic, evolved, singular entity—a social organism. The social organism subordinated its constituent individuals, and its health, welfare, and morals trumped the individual's rights and liberties.

Progressivism reconstructed American liberalism by dismantling the free market of classical liberalism and erecting in its place the welfare state of modern liberalism. The new liberalism discarded economic liberties as archaic impediments to necessary improvements to society's health, welfare, and morals.

It is well known that modern liberalism permanently demoted economic liberties. Few twenty-first-century progressives think that minimum wages or maximum hours or occupational licensing unjustly infringe upon a worker's right to freely contract on her own behalf.

But the original progressives' illiberal turn did not stop at property and contract rights. They assaulted political and civil liberties, too, trampling on individual rights to person, to free movement, to free expression, to marriage and to reproduction. The progressives denied millions these basic freedoms, on grounds that their inferiority threatened America's economic and hereditary security. They were wrong on both counts. That did not stop them, nor has it stopped those who, unaware of the history, repeat the same false claims today.

NOTES

PROLOGUE

1. Fourth branch agencies were chartered to be independent of their creators, and they were, uniquely, endowed with legislative, executive, and judicial powers combined. Agencies make law—regulatory rules have the full power of federal legislation—they enforce regulation, and they adjudicate regulatory disputes.

2. Richard Dugdale (1877) *The Jukes: A Study in Crime, Pauperism, Disease, and Heredity*. New York: G. P. Putnam's Sons.

CHAPTER 1: REDEEMING AMERICAN ECONOMIC LIFE

1. Stanley Lebergott (1966) "Labor Force and Employment, 1800–1960." In Dorothy Brady (ed.), *Output, Employment, and Productivity in the United States after 1800*. Cambridge, MA: National Bureau of Economic Research, pp. 117–204, p. 119.

2. US Bureau of the Census (1975) *Historical Statistics of the United States*, Vol. I. Series C 89–119: Immigrants, by country, 1820–1970, Washington, DC: Government Printing Office, pp. 105–6. Niles Carpenter (1927) *Immigrants and Their Children, 1920*. US Census Monographs VII. Washington, DC: Government Printing Office, p. 27. *Historical Statistics of the United States*. Cambridge: Cambridge University Press, Table Ba1131–1144; Table Ba1145–1158. Available at http://hsus.cambridge.org/HSUSWeb/toc/hsusHome.do.

3. Gerald Friedman (1999) "U.S. Historical Statistics: New Estimates of Union Membership in the United States, 1880–1914." *Historical Methods: A Journal of Quantitative and Interdisciplinary History* 32(2): 75–86.

4. This figure includes strikes only and omits employer lockouts, slowdowns, and other types of labor disputes. Union-organized strikes comprise about two-thirds of the total strikes. US Department of Labor (1906) *Twenty-First Annual Report of the Commissioner of Labor*. Washington, DC: Government Printing Office, pp. 15–16.

5. Naomi R. Lamoreaux (1988) *The Great Merger Movement in American Business, 1895–1904*. Cambridge: Cambridge University Press, pp. 1–2.

6. Thomas McCraw (1984) "Business and Government: The Origins of an Adversary Relationship." *California Management Review* 26: 33–52, p. 42, citing Alfred Chandler, Jr. (1977) *The Visible Hand: The Managerial Revolution in American Business*. Cambridge, MA: Harvard University Press.

7. Thomas J. Haskell (1977) *The Emergence of Professional Social Science: The American Social Science Association and the Nineteenth Century Crisis of Authority*. Urbana, IL: University of Illinois Press, p. 1.

8. Simon Nelson Patten (1907) *The New Basis of Civilization.* New York: Macmillan. Richard T. Ely (1910) "The American Economic Association 1885–1909." *American Economic Association Quarterly,* 3rd Series, 11(1): 47–111, p. 67. Frederick Jackson Turner (1911) "Social Forces in American History." *American Historical Review* 16(2): 217–33. John Dewey (1915 [1899]) *The School and Society,* revised ed. Chicago: University of Chicago Press, p. 6.

9. David A. Wells (1890) *Recent Economic Changes and Their Effect on the Production and Distribution of Wealth and the Well-being of Society.* New York: D. Appleton and Co., p. v.

10. Benjamin P. DeWitt (1915) *The Progressive Movement: A Non-partisan, Comprehensive Discussion of Current Tendencies in American Politics.* New York: Macmillan.

11. Steven J. Diner (1999) "Linking Politics and People: The Historiography of the Progressive Era." *OAH Magazine of History* 13(2): 5–9.

12. Henry May (1949) *Protestant Churches and Industrial America.* New York: Harper Bros., p. 235.

13. "Strange is it not?" Ely wrote of the Knights of Labor, "that the despised trades-union and labor organizations should have been chosen to perform this high duty of conciliation! But hath not God ever called the lowly to the most exalted missions, and hath not he ever called the foolish to confound the wise?" Richard T. Ely (1886) *The Labor Movement in America.* New York: Thomas Crowell and Company, p. 139.

14. Linda Gordon (2002) "SHGAPE Distinguished Historian Address: If the Progressives Were Advising Us Today, Should We Listen?" *Journal of the Gilded Age and Progressive Era* 1(2): 109–21, p. 111.

15. Ely accused La Follette of having been "more help to the Kaiser than a quarter of a million troops." Carol Gruber (1975) *Mars and Minerva: World War I and the Uses of Higher Learning in America.* Baton Rouge: Lousiana State University Press, p. 208.

16. DeWitt, *The Progressive Movement,* p. viii.

17. Daniel Rodgers (1982) "In Search of Progressivism." *Reviews in American History* 10(4): 113–32.

18. Herbert Croly (1911 [1909]) *The Promise of American Life.* New York: Macmillan, p. 49.

19. Eldon Eisenach (1994) *The Lost Promise of Progressivism.* Lawrence: University of Kansas Press, p. 5.

20. Rodgers, "In Search of Progressivism," pp. 123, 126.

21. Richard T. Ely (1886) "Report on the Organization of the American Economic Association." *Publications of the American Economic Association* 1(1): 5–16, pp. 6–7.

22. Robert Wiebe (1967) *The Search for Order, 1877–1920.* New York: Hill and Wang, p. 166.

23. When Walter Lippmann quipped in 1921, "an American will endure almost any insult except the charge that he is not progressive," he recognized that *progressive* was a term of approbation, connoting such virtues as goodness, far-sightedness, or enlightenment. Progressivism's critics did not oppose these virtues or the benefits of progress; their disagreement concerned how progress was best attained. This difficulty is one of many that make "progressives" a contested name for Progressive Era reformers. Some historians have preferred to call the progressives "new liberals" in a nod to their English confreres and to emphasize the historical continuities with and departures from classical liberalism. Progressive Era historiography gives us an embarrassment of terminological diversity: "progressive liberals," "democratic lib-

erals," "welfare state liberals," "democratic collectivists," and "social democrats," among other variants. This book uses the old-fashioned "progressive," because, whatever its shortcomings, it refers to specifically to American reform and is inescapably embedded in the language of contemporaries and in the writing of historians of the Progressive Era. [Richard L. McCormick (1997) "Public Life in Industrial America, 1877–1917." In Eric Foner (ed.), *The New American History*. Philadelphia: Temple University Press: 93–117, p. 121.]

24. In the nineteenth-century American context, "evangelical" refers less to a specific group or church than to a particular set of Protestant religious commitments: (1) conversionism, the need to be "born again"; (2) biblicism, the authority of the bible; (3) crucicentrism, an emphasis on the sacrifice and resurrection of Jesus Christ; and (4) activism, in the form of missionary or social work [David W. Bebbington (1989) *Evangelicalism in Modern Britain: A History from the 1730s to the 1980s*. London: Unwin, Hyman, pp. 2–3].

25. Arthur Schlesinger (1932) "A Critical Period in American Religion, 1875–1900." *Proceedings of the Massachusetts Historical Society* 64: 523–47.

26. "The laws of exchange are based on nothing less solid than the will of God," Perry wrote. Arthur Latham Perry (1875) *The Elements of Political Economy*. New York: Scribner, p. 105.

27. Perry, *The Elements of Political Economy*, p. 166.

28. Bradley Bateman has recovered the influence of the social gospel movement on the progressive economists who founded the AEA. See, for example, Bradley W. Bateman (1998) "Clearing the Ground: The Demise of the Social Gospel Movement and the Rise of Neoclassicism in American Economics." In Mary S. Morgan and Malcolm Rutherford (eds.), *From Interwar Pluralism to Postwar Neoclassicism: History of Political Economy* 30 (supplement): 29–52; Bradley W. Bateman (2001) "Make a Righteous Number: Social Surveys, the Men and Religion Forward Movement, and Quantification in American Economics." In Judy L. Klein and Mary S. Morgan (eds.), *The Age of Economic Measurement: History of Political Economy* 33 (supplement): 57–85.

29. Josiah Strong (1885) *Our Country: Its Possible Future and Its Present Crisis*. New York: Baker & Taylor.

30. A. W. Coats (1988) "The Educational Revolution and the Professionalization of American Economics." In William J. Barber (ed.), *Breaking the Academic Mould: Economists and American Higher Learning in the Nineteenth Century*. Middletown, CT: Wesleyan University Press, pp. 340–75.

31. Richard T. Ely (1891) "Pauperism in the United States." *North American Review* 152(413): 395–409. Quote is from pp. 406–7.

32. Richard T. Ely (1889) *Social Aspects of Christianity*. New York: Thomas Crowell and Company, p. 53.

33. John R. Commons (1894) *Social Reform and the Church*. New York: Thomas Crowell and Company, p. 71.

34. Edward A. Ross (1907) *Sin and Society: An Analysis of Latter-Day Iniquity*. New York: Macmillan. Though Ross revisited the Christian idea of inborn sinfulness, he did not believe that all human beings were perfectible by good works. On the contrary, as a eugenicist, Ross merely substituted biological inadequacy for spiritual inadequacy. For Ross, persons from racially inferior groups were, if not fallen as such, nonetheless hereditarily beyond uplift.

35. Eisenach, *The Lost Promise of Progressivism*, p. 11.

36. Richard T. Ely (1896) *The Social Law of Service*. New York: Eaton & Mains, pp. 162–63.

37. Commons, *Social Reform and the Church*, p. 54.

38. Joseph Dorfman (1969 [1954]) "Introduction." In Joseph Dorfman (ed.), *Two Essays by Henry Carter Adams*. New York: Augustus M. Kelley, p. 9. See also A. W. Coats (1968) "Henry Carter Adams: A Case Study in the Emergence of the Social Sciences in the United States, 1850–1900." *Journal of American Studies* 2(2): 177–97.

39. Cited in Ajay K. Mehrotra (2013) *Making the Modern American Fiscal State: Law, Politics, and the Rise of Progressive Taxation, 1877–1929*. Cambridge: Cambridge University Press, p. 100.

40. Edwin R. A. Seligman (1922) "Memorial to Former President Henry C. Adams." *American Economic Review* 12(3): 401–8, p. 403.

41. John Henry (1995) *John Bates Clark: The Making of a Neoclassical Economist*. London: Macmillan, pp. 1–2.

42. John R. Commons (1934) *Myself*. New York: Macmillan, pp. 8, 43–44.

43. In 1892, Small founded and chaired the Department of Sociology, which was colocated in the Chicago Divinity School. Small was the founding editor of the *American Journal of Sociology* (1895), which he edited for many years.

44. Cited in Steven Diner (1975) "Department and Discipline: The Department of Sociology at the University of Chicago, 1892–1920." *Minerva* 13(4): 514–53, p. 524.

45. Jane Addams (2002 [1893]) "The Subjective Necessity for Social Settlements." Reprinted in Jean Bethke Elshtain (ed.), *Jane Addams and the Dream of American Democracy*. New York: Basic Books.

46. Ely, *Introduction to Political Economy*, p. 118.

47. In Mary Furner's formulation, objectivity required avoiding too much advocacy. Mary Furner (1975) *Advocacy and Objectivity: A Crisis in the Professionalization of American Social Science 1865–1905*. Lexington: University Press of Kentucky.

48. Delegates attending the 1912 Progressive Party convention, which nominated Theodore Roosevelt for president, sang rousing choruses of "Onward Christian Soldiers."

49. David Hollinger (1989) "Justification by Verification: The Scientific Challenge to the Moral Authority of Christianity in Modern America." In Michael J. Lacey (ed.), *Religion and Twentieth-Century American Intellectual Life*. Cambridge: Cambridge University Press, pp. 116–35, pp. 117, 123.

CHAPTER 2: TURNING ILLIBERAL

1. Joseph Dorfman (1955) "The Role of the German Historical School in American Economic Thought." *American Economic Review* 55(2): 17–28. Jurgen Herbst (1965) *The German Historical School in American Scholarship: A Study in the Transfer of Culture*. Ithaca, NY: Cornell University Press. Axel Schäfer (2000) *American Progressives and German Social Reform 1875–1920*. Stuttgart: Franz Steiner Verlag. Daniel Rodgers (1998) *Atlantic Crossings: Social Politics in a Progressive Age*. Cambridge, MA: Harvard University Press, pp. 76–111.

2. In 1906, Henry Farnam of Yale University polled America's leading economists to ascertain the influence of German economic ideas. Of the 116 who replied, 59 had studied in Germany, and a larger number acknowledged influence. Of several influential German pro-

fessors, Wagner and Schmoller had the most American students. Farnam was a student of Schmoller's. See Benny Carlson (1999) "Wagner's Legacy in America: Re-opening Farnam's Inquiry." *Journal of the History of Economic Thought* 21(3): 289–310.

3. Richard T. Ely (1884) "The Past and Present of Political Economy." In Herbert B. Adams (ed.), *Johns Hopkins University Studies in Historical and Political Science*, Vol II, no. 3 Baltimore: Johns Hopkins University, p. 64.

4. Richard T. Ely (1883) *French and German Socialism in Modern Times*. New York: Harper Bros., pp. 236–37.

5. Fritz Ringer (1969) *Decline of the German Mandarins: The German Academic Community, 1890–1933*. Cambridge, MA: Harvard University Press.

6. Nicholas W. Balabkins (1988) *Not by Theory Alone ... The Economics of Gustav von Schmoller and Its Legacy to America*. Berlin: Dunker & Humboldt, p. 38.

7. F. W. Taussig (1885) "College Graduates in Germany." *The Nation*, April 2, p. 275. Cited in Dorfman,"The Role of the German Historical School in American Economic Thought," p. 22.

8. John B. Parrish (1967) "The Rise of Economics as an Academic Discipline: The Formative Years to 1900." *Southern Economic Journal* 34 (July): 1–16, p. 11.

9. A. W. Coats (1988) "The Educational Revolution and the Professionalization of American Economics." In William J. Barber (ed.), *Breaking the Academic Mould: Economists and American Higher Learning in the Nineteenth Century*. Middletown, CT: Wesleyan University Press, pp. 340–75, pp. 342–43.

10. Alexandra Oleson and J. Voss (eds.) (1979) *Organization of Knowledge in Modern America, 1860–1920*. Baltimore: Johns Hopkins University Press, p. xii.

11. William James (1903) "The PhD Octopus." *Harvard Monthly*, March.

12. Parrish, "The Rise of Economics as an Academic Discipline," pp. 9–10.

13. Parrish, "The Rise of Economics as an Academic Discipline," pp. 9–10.

14. Thomas J. Haskell (1977) *The Emergence of Professional Social Science: The American Social Science Association and the Nineteenth Century Crisis of Authority*. Urbana: University of Illinois Press, p 24.

15. Harvard's *Quarterly Journal of Economics* and Columbia's *Political Science Quarterly* appeared in 1886, the same year the AEA began publishing its monograph series. Pennsylvania's *Annals of the American Academy of Political and Social Science* began publishing in 1890. *The Yale Review* commenced in 1892, as did Chicago's *Journal of Political Economy*, which appeared before the University had been in operation for three months.

16. Ely produced several textbooks. His *Outlines of Economics* was first published in 1893. Three coauthors joined for the subsequent four revised editions—Thomas S. Adams, Max O. Lorenz, and Allyn A. Young—published over thirty years. *Outlines* began as a revision of Ely's *Introduction*, but evolved into a distinct book. The *Introduction* was quite successful, in part because Ely sold the book at Chautauqua and other camp meetings. One estimate put total sales at 500,000 copies. [Sidney Fine (1959) *Laissez Faire and the General-Welfare State*. Ann Arbor: University of Michigan Press: pp. 238–39.]

17. A. W. Coats (1985) "The American Economic Association and the Economics Profession." *Journal of Economic Literature* 23(4): 1697–1727, p. 1699.

18. Charles K. Adams was president of Cornell when the AEA organized in 1885, later leading Wisconsin. Francis Amasa Walker presided over MIT from 1881 to 1897. E. Benjamin Andrews led Brown (1889–1898) and the University of Nebraska (1900–1908), and

Arthur T. Hadley led Yale from 1899 to 1921. Woodrow Wilson was Princeton's president from 1902 to 1910. Carroll Wright led Clark University from 1902 to 1909. After heading the University of Pennsylvania's Wharton School, Edmund J. James became president of Northwestern University (1902) and the University of Illinois (1904–1920).

19. This point is due to Roger L. Geiger (2004) *To Advance Knowledge: The Growth of the American Research Universities, 1900–1940*. New Brunswick, NJ: Transaction Publishers, p. 21.

20. Fetter instanced "J. B. Clark, E. J. James, H. W. Farnam, S. N. Patten, R. T. Ely, A. T. Hadley, E. R. A. Seligman, A. W. Small, and F. W. Taussig; in the next decade, W. M. Daniels, H. B. Gardner, E. F. Gay, and others." "When can we hope," Fetter wondered, "that among a thousand [PhDs in economics] will be found another such Hall of Fame?" Frank A. Fetter (1925) "The Economists and the Public." *American Economic Review* 15(1): 13–26, p. 14.

21. Richard T. Ely (1886) "Report on the Organization of the American Economic Association." *Publications of the American Economic Association* 1(1): 5–16, p. 20, note 1.

22. Ely, "Past and Present of Political Economy," p. 50.

23. Cited in Michael McGerr (2003) *A Fierce Discontent: The Rise and Fall of the Progressive Movement in America, 1870–1920*. New York: Free Press, p. 217.

24. John R. Commons (2009 [1934]) *Institutional Economics: Its Place in Political Economy*, vol. 2, with an Introduction by Malcolm Rutherford. New Brunswick, NJ: Transaction Publishers, p. 842.

25. Daniel Rodgers (1987) *Contested Truths: Keywords in American Politics Since Independence*. New York: Basic Books, p. 179. Edward A. Ross (1896) "Social Control." *American Journal of Sociology* 1(5): 513–35, p. 518.

26. Edward A. Ross (1901) *Social Control: A Survey of the Foundations of Order*. New York: Macmillan, p. 168.

27. Leon Fink (1997) *Progressive Intellectuals and the Dilemmas of Democratic Commitment*. Cambridge, MA: Harvard University Press, p. 17.

28. Edward A. Ross (1936) *Seventy Years of It*. New York: D. Appleton-Century, p. 42.

29. Henry Steele Commager (ed.) (1967) *Lester Ward and the American Welfare State*. Indianapolis, IN: Bobbs-Merrill. Samuel Chugerman (1939) *Lester Frank Ward: The American Aristotle*. Durham, NC: Duke University Press. Some historians are skeptical of Ward's influence. See Hamilton Cravens (1978) *The Triumph of Evolution: American Scientists and the Heredity-Environment Controversy 1900–1941*. Philadelphia: University of Pennsylvania Press, pp. 136–37.

30. Lester Frank Ward (1883) *Dynamic Sociology or Applied Social Science as Based upon Statical Sociology and the Less Complex Sciences*. New York: D. Appleton and Co.

31. Laura L. Lovett (2007) *Conceiving the Future*. Chapel Hill: University of North Carolina Press, p. 81.

32. Ross, *Social Control*, p. 74.

33. Cited in David Greenberg (2011) "Beyond the Bully Pulpit." *Wilson Quarterly* 35(3): 22–29, p. 27.

34. Henry Carter Adams (1886) "Economics and Jurisprudence." *Science* 8 (178): 15–19, p. 17. Herbert Croly (1911 [1909]) *The Promise of American Life*. New York: Macmillan, p. 414. Edward A. Ross (1901) *Social Control: A Survey of the Foundations of Order*. New York: Macmillan, p. 67. Richard T. Ely (1889) *Introduction to Political Economy*. New York: Chautauqua Press, p. 92.

35. *New Republic* (1915) "The Bill of Rights Again." *New Republic*, April 17, pp. 272–73.

36. Woodrow Wilson (1908) *Constitutional Government in the United States*. New York: Columbia University Press, p. 16, cited in David Bernstein (2011) *Rehabilitating Lochner*. Chicago: University of Chicago Press, p. 92.

37. Cited in Morton Keller (1994) *Regulating a New Society: Public Policy and Social Change in America, 1900–1933*. Cambridge, MA: Harvard University Press, p. 80.

38. Charles A. Beard (1921[1913]) *An Economic Interpretation of the Constitution of the United States*. New York: Macmillan.

39. Daniel T. Rodgers (2008) "Rights Consciousness in American History." In David J. Bodenhamer and James W. Ely, Jr. (eds.), *The Bill of Rights in Modern America*, revised ed. Bloomington: Indiana University Press, p. 18.

40. *Santa Clara County v. Southern Pacific Railroad Company*, 118 US 394 (1886).

41. Richard T. Ely (1915) "Progressivism True and False—An Outline." *American Review of Reviews* 51: 209–11, p. 209.

42. This formulation of the problem is due to Richard P. Adelstein (1991) " 'The Nation as an Economic Unit': Keynes, Roosevelt, and the Managerial Ideal." *Journal of American History* 78(1): 160–87, p. 166.

CHAPTER 3: BECOMING EXPERTS

1. See Theda Skocpol (1992) *Protecting Soldiers and Mothers: The Political Origins of Social Policy in the United States*. Cambridge, MA: Belknap Press of Harvard University Press.

2. Richard Adelstein (1988) "Mind and Heart: Economics and Engineering at MIT." In William J. Barber (ed.), *Breaking the Academic Mould: Economists and American Higher Learning in the Nineteenth Century*. Middletown, CT: Wesleyan University Press, pp. 290–317.

3. Louis Menand (2001) *The Metaphysical Club*. New York: Farrar, Straus, Giroux, p. 302.

4. W. W. Folwell (1893) "The New Economics." *Publications of the American Economic Association* 8(1): 19–40.

5. Francis Amasa Walker (1889) "The Recent Progress of Political Economy in the United States." *Publications of the American Economic Association* 4(4): 17–40, p. 29.

6. US Bureau of the Census (1975) *Historical Statistics of the United States, Colonial Times to 1970*, vol. I. series D 85-86. Washington, DC: Government Printing Office, p. 135. Romer's revision produces lower peak figures, but her unemployment rate is still above 11 percent from 1893 to 1897. Christina Romer (1986) "Spurious Volatility in Historical Unemployment Data." *Journal of Political Economy* 94(1): 1–37, p. 31.

7. Matthew Algeo (2011) *The President Is a Sick Man*. Chicago: Chicago Review Press.

8. Arthur T. Hadley (1899) "The Relationship between Economics and Politics." *Yale Law Journal* 8(4): 194–206, p. 194.

9. Henry Rand Hatfield (1899) "The Chicago Trust Conference." *Journal of Political Economy* 8(1): 1–18, pp. 17–18. I thank Nicola Giocoli for the reference.

10. Edwin R. A. Seligman (1903) "Economics and Social Progress." *Publications of the American Economic Association*, third series, 4(1): 52–70, p. 69.

11. Arthur T. Hadley (1900) "Economic Theory and Political Morality" *Publications of the American Economic Association*, 3rd Series, Vol. 1, No. 1, *Papers and Proceedings of the Twelfth Annual Meeting, Ithaca, N. Y., December 27–29, 1899:* 45–61, p. 61.

12. John R. Commons, Edwin R. A. Seligman, L. M. Keasbey, E. W. Bemis, Professor Mayo-Smith, and Professor Hadley (1900) "Discussion of the President's Address."*Publications of the American Economic Association*, third series 1(1): 62–88, 287–88.

13. Commons et al., "Discussion of the President's Address."

14. Steven G. Medema (2007) "The Hesitant Hand: Mill, Sidgwick and the Evolution of the Theory of Market Failure." *History of Political Economy* 39(3): 331–58.

15. John Stuart Mill (1848) *Principles of Political Economy: With Some of Their Applications to Social Philosophy*, vol. II. London: John W. Parker, pp. 515–16.

16. John Maynard Keynes (1926) *The End of Laissez-Faire*. London: Hogarth Press.

17. Richard T. Ely (1889) *Introduction to Political Economy*. New York: Chautauqua Press, p. 38.

18. Hadley, "The Relation between Economics and Politics," p. 194.

19. Seligman, "Economics and Social Progress," p. 70.

20. David Moss (1996) *Socializing Security: Progressive-Era Economists and the Origins of American Social Policy*. Cambridge, MA: Harvard University Press, p. 16.

21. Moss, *Socializing Security*, p. 17.

22. Dorothy Ross (1984) "American Social Science and the Idea of Progress." In Thomas L. Haskell (ed.), *The Authority of Experts*. Bloomington: Indiana University Press, pp. 157–75, p. 157.

23. Richard T. Ely (1903) *Studies in the Evolution of Industrial Society*. New York: Macmillan, p. 456.

24. Edward A. Ross (1907) *Sin and Society: An Analysis of Latter-Day Iniquity*. New York: Macmillan, p. 41.

25. Edward A. Ross (1916) "Conscience of the Expert." *School & Society* 3: 522–24.

26. Daniel Rodgers (1982) "In Search of Progressivism." *Reviews in American History* 10(4): 113–32, p. 125.

27. Daniel Rodgers (1998) *Atlantic Crossings*. Cambridge, MA: Harvard University Press, p. 236.

28. Mary Furner (1975) *Advocacy and Objectivity: A Crisis in the Professionalization of American Social Science 1865–1905*. Lexington: University Press of Kentucky.

29. Adna F. Weber (1907) "Labor Legislation, National and International." *Journal of Social Science* 45: 36.

30. Moss, *Socializing Security*, p. 32.

31. Josephine Goldmark (1953) *Impatient Crusader*. Urbana: University of Illinois Press, p. 69.

32. John R. Commons (1913) *Labor and Administration*. New York: Macmillan, p. 7.

33. Commons, *Labor and Administration*, pp. 7–13.

34. Quoted in Samuel Haber (1988) "Expertise." In John D. Buenker and Edward R. Kantowicz (eds.), *Historical Dictionary of the Progressive Era, 1890–1920*. Westport, CT: Greenwood Press, pp. 130–31.

35. Samuel Haber (1964) *Efficiency and Uplift: Scientific Management in the Progressive Era*. Chicago: University of Chicago Press.

36. Thorstein Veblen (1906) "The Place of Science in Modern Civilization." *American Journal of Sociology* 11(5): 585–609, p. 600.

37. *Adkins v. Children's Hospital*, 261 US 525 (1923) (USSC+).

38. Quoted in Julie A. Reuben (1996) *The Making of The Modern University*. Chicago: University of Chicago Press, p. 152.

39. Theodore Roosevelt (1911) "The Trusts, the People and the Square Deal." *Outlook* 99(12): 649–56, p. 654.

40. See especially Herbert Croly (1911 [1909]) *The Promise of American Life*. New York: Macmillan pp. 169–70.

41. William G. Sumner (1911) "The Conquest of the United States by Spain." In A. G. Keller (ed.), *War and Other Essays by William Graham Sumner*. New Haven, CT: Yale University Press, pp. 297–334, p. 334.

42. Carol Gruber (1975) *Mars and Minerva: World War I and the Use of the Higher Learning in America*. Baton Rouge: Louisiana State University Press.

43. William Graham Sumner (1924) "Democracy and Plutocracy." In A. G. Keller and M. R. Davie (eds.), *Selected Essays of William Graham Sumner*. New Haven, CT: Yale University Press, p. 214. Sumner, *War and Other Essays*, p. 160.

44. William Graham Sumner (1906) *Folkways*. Boston: Ginn and Co., p. 170.

45. For a history of Wisconsin Institutional economics that emphasizes the centrality of Commons, see Malcolm Rutherford (2006) "Wisconsin Institutionalism: John R. Commons and His Students." *Labor History* 47(2): 161–88.

46. McCarthy directed the Wisconsin Legislative Reference Library, known colloquially as the "bill factory," which provided research assistance and model legislation for the economists and other University experts fashioning scientific legislation in Madison.

47. *Manitowoc Daily Herald* (1913) "The Wisconsin Idea Told by Pres. Van Hise." *Manitowoc Daily Herald*, June 5, p. 8.

48. Charles McCarthy (1912) *The Wisconsin Idea*. With an introduction by Theodore Roosevelt. New York: Macmillan, pp. 28, 29.

49. John Dewey (1915) *German Philosophy and Politics*. New York: Henry Holt, pp. 15–16.

50. Frederic C. Howe (1912) *Wisconsin: An Experiment in Democracy*. New York: Charles Scribner's Sons, pp. vii, 38, 189, 174.

51. Charles Van Hise (1913) "Address by the President of the University of Wisconsin." *City Club of Philadelphia Bulletin* 6(20): 469–75, p. 473.

52. The US president has the power to appoint commissioners or board members, but they must be drawn from both parties, and their terms are longer than the president's and also are staggered so as to deter "packing." A president cannot remove a commissioner or board member, except for cause. Congress also cannot remove commissioners from those agencies with executive authority, nor can members of Congress serve on them. Other federal agencies of the fourth branch agencies are less protected, but they too were created to be insulated from politics, and they too employ a permanent, professional civil service that vastly outnumbers political appointees (about one hundred to one in the twenty-first century). Federal agencies likewise enjoy broad discretionary powers to surveil, investigate, and regulate, and their regulations also have the full power of federal law. The term "fourth branch" is somewhat anachronistic when applied to the Progressive Era, but the concept and its realization in the form of the administrative state are both quintessential products of the period.

53. Cited in Ajay Mehrotra (2004) "Envisioning the Modern American Fiscal State: Progressive Era Economists and the Intellectual Foundations of the U.S. Income Tax." *UCLA Law Review* 54: 1793–1866.

54. For an incisive chronicle of political economists' role in laying the foundations of the income tax, see Ajay Mehrotra (2013) *Making the Modern American Fiscal State: Law, Politics, and the Rise of Progressive Taxation, 1877–1929.* New York: Cambridge University Press.

55. Edwin R. A. Seligman (1895) *Essays in Taxation.* New York: Macmillan, p. 72.

56. Edwin R. A. Seligman (1914) *The Income Tax: A Study of the History, Theory, and Practice of Income Taxation at Home and Abroad,* second ed. New York: Macmillan, p. 4.

57. The corporate tax law of 1909 was superseded by the Corporation Tax Act of September 8, 1916, which taxed corporate income. The Inheritance Tax Act was passed the same day.

58. The Department of Commerce and Labor was created on February 14, 1903. On March 4, 1913, Congress split it into the Department of Commerce and the Department of Labor.

59. Clayton Act, October 15, 1914.

60. Lloyd-LaFollette Act, August 24, 1912, 37 Stat. 555, ch. 389.

61. Eight Hours on Public Works Act, March 3, 1913; Eight Hour Day Act (also known as the Adamson Act), September 3 and 5, 1916.

62. Compensation of Injured Federal Employees Act, September 7, 1916.

63. Mann-Elkins Act, June 18, 1910.

64. Farm Loan Act, July 17, 1916.

65. The White Slave-Traffic Act (also known as the Mann Act), June 25, 1910.

66. Food and Drug Act, June 30, 1906.

67. Meat Inspection Act, March 4, 1907.

68. Sale and Use of Cocaine and Narcotics Act, December 17, 1914; Opium Act, January 17, 1914.

69. Children's Bureau Act, April 9, 1912.

70. Child Labor Act, September 1, 1916.

71. Vocational Education Act, December 23, 1917. This legislation was passed after US entry into the First World War.

72. Hepburn Act, June 29, 1906.

73. Mediation and Arbitration Law Act, July 15, 1913.

74. Elkins Act, February 19 1903.

75. Shipping Board Act, September 7, 1916.

76. Postal Savings Act, June 25, 1910.

77. Acts of 1901–1902, fifty-seventh Congress, first session, chapter 641.

78. Immigration Act, February 5, 1917.

79. J. Stanley Lemons (1988) "Mothers' Pensions Acts." In John D. Buenker and Edward R. Katowicz (eds.), *Historical Dictionary of the Progressive Era, 1890–1920.* Westport, CT: Greenwood Press, pp. 291–92.

80. Elizabeth Brandeis (1966 [1935]) "Labor Legislation." In *History of Labor in the United States, 1896–1932,* vol. III. New York: Augustus M. Kelley, pp. 399–700, pp. 459, 501, ff. 1.

81. William J. Novak (2008) "The Myth of the 'Weak' American State." *American Historical Review* 113(3): 752–72.

82. Morton Keller (1990) *Regulating a New Economy, Public Policy and Economic Change in America, 1900–1933.* Cambridge, MA: Harvard University Press, p. 88.

83. W. Elliot Brownlee (1990) "Economists and the Formation of the Modern Tax System in the United States: The World War I Crisis." In Mary O. Furner and Barry E. Supple (eds.), *The State and Economic Knowledge: The American and British Experiences*. Cambridge: Cambridge University Press, pp. 401–35.

84. Edwin R. A. Seligman (1921) *Essays in Taxation*. New York: Macmillan, p. 679, 694.

85. Roy G. Blakey (1917) "The War Revenue Act of 1917." *American Economic Review* 7(4): 791–815, p. 791.

86. In 1916, US government expenditures were $734 million. In 1922, they were $3,372 million. Adjusted for inflation (47 percent for 1916–1922), US government expenditures more than tripled. Even when not counting US debt service costs, government spending more than doubled. US Bureau of the Census (1949) *Historical Statistics of the United States, 1789–1945*. Series P 99-108. Washington, DC: Government Printing Office, p. 299.

87. John A. Lapp (1917) *Important Federal Laws*. Indianapolis, IN: B.F. Bowen & Co, p. iv.

88. Brownlee, "Economists and the Formation of the Modern Tax System in the United States," p. 407.

89. Robert William Fogel, Enid M. Fogel, Mark Guglielmo, and Nathaniel Grotte (2013) *Political Arithmetic: Simon Kuznets and the Empirical Tradition in Economics*. Chicago: University of Chicago Press, pp. 30–32.

90. Quoted in Ray Stannard Baker (1911) "The Gospel of Efficiency: A New Science of Business Management." *American Magazine*, March, p. 563.

91. Robert Cuff (1989) "Creating Control Systems: Edwin F. Gay and the Central Bureau of Planning and Statistics, 1917–19." *Business History Review* 63: 588–613.

92. Quoted in Michael McGerr (2003) *A Fierce Discontent: The Rise and Fall of the Progressive Movement in America, 1870–1920*. New York: Free Press, pp. 285, 299.

93. John Dewey (1929 [1918]) "Propaganda." In Joseph Ratner (ed.), *Characters and Events: Popular Essays in Social and Political Philosophy*, vol. 2. New York: Henry Holt, p. 517. John Dewey (1929 [1918]) "The Social Possibilities of War." In Ratner (ed.), *Characters and Events*, vol. 2, p. 557.

94. Wesley Claire Mitchell (1924) "The Prospects of Economics." In Morris A. Copeland (ed.), *The Trend of Economics*. New York: Knopf, p. 33.

95. Benjamin P. DeWitt (1915) *The Progressive Movement: A Non-partisan, Comprehensive Discussion of Current Tendencies in American Politics*. New York: Macmillan, pp. 6–7.

96. Richard L. McCormick (1981) "The Discovery That Business Corrupts Politics: A Reappraisal of the Origins of Progressivism." *American Historical Review* 86(2): 257–74.

97. Mark Lawrence Kornbluh (2000) *Why America Stopped Voting: The Decline of Participatory Democracy and the Emergence of Modern American Politics*. New York: New York University Press.

98. Woodrow Wilson (1901) "The Reconstruction of the Southern States." *Atlantic Monthly*, January, pp. 1–15, quotes from p. 6. Eric Foner (1998) *The Story of American Freedom*. New York: W.W. Norton & Co., p. 186.

99. John R. Commons (1907) *Races and Immigrants*. New York: Macmillan, p. 41.

100. Edward A. Ross (1912) *Changing America: Studies in Contemporary Society*. New York: Century, pp. 4–5.

101. Foner, *The Story of American Freedom*, pp. 185–86.

102. These figures are from Robert Wiebe (1995) *Self-Rule: A Cultural History of American Democracy*. Chicago: University of Chicago Press, pp. 134–35.

103. James Morone referred to this tension as the "progressive oxymoron." James Morone (1998) *The Democratic Wish: Popular Participation and the Limits of American Government.* New Haven, CT: Yale University Press, pp. 98, 126.

104. Lenore O'Boyle (1983) "Learning for Its Own Sake: The German University as Nineteenth-Century Model." *Comparative Study of Society and History* 25(1): 3–25, pp. 23–24.

105. Lester Frank Ward (1894 [1883]) *Dynamic Sociology*, vol. I. New York: Appleton, p. 37.

106. Walter Lippmann (1922) *Public Opinion*. New York: Harcourt Brace & Co.

107. Albert Beveridge (1908) *The Meaning of the Times: And Other Speeches*. Indianapolis: Bobbs Merrill, p. 49. Commons, *Races and Immigrants,* p. 3.

108. Richard T. Ely (1898) "Fraternalism vs. Paternalism in Government." *Century Magazine* 55(5): 780–85, p. 781. Edward A. Ross (1901) *Social Control: A Survey of the Foundations of Order*. New York: Macmillan, pp. 83–84. Edward A. Ross (1920) *Principles of Sociology*. New York: Century, p. 268.

109. Ely, *Studies in the Evolution of Industrial Society*, p. 456. Richard T. Ely (1894) "The Church and the Labor Movement." *Outlook*, January 13, pp. 60, 62.

110. Richard T. Ely (1882) "Administration of the City of Berlin." *Nation,* March 23. pp. 245–46, p. 246. Cited in Axel R. Schäfer (2000) *American Progressives and German Social Reform, 1875–1920*. Stuttgart: Franz Steiner Verlag, p. 86.

111. Richard T. Ely (1922) "The Price of Progress." *Administration* 3(6): 657–63, p. 661.

112. Irving Fisher (1907) "Why Has the Doctrine of Laissez Faire Been Abandoned?" *Science* 25(627): 18–27.

113. Fisher, "Why Has the Doctrine of Laissez Faire Been Abandoned?" pp. 19–21.

114. Foner, *The Story of American Freedom*, pp. 154–55.

115. Cited in Leon Fink (1997) *Progressive Intellectuals and the Dilemmas of Democratic Commitment*. Cambridge, MA: Harvard University Press, p. 14. I am indebted to Fink's subtle discussion of these issues.

116. Edwin R. A. Seligman (1899) "Discussions of the President's Address." *Publications of the American Economic Association*, third series, 1(1): 62–88, 287–88.

CHAPTER 4: EFFICIENCY IN BUSINESS AND PUBLIC ADMINISTRATION

1. Samuel Haber (1988) "Efficiency." In John D. Buenker and Edward R. Kantowicz (eds.), *Historical Dictionary of the Progressive Era 1890–1920.* Westport, CT: Greenwood Press, pp. 130–31.

2. Jane Addams (1910) "Charity and Social Justice." *North American Review* 192(656): 68–81, p. 79. Helen Sumner (1910) "The Historical Development of Women's Work in the United States." *Proceedings of the Academy of Political Science in the City of New York* 1(1): 11–26, p. 26.

3. Naomi Lamoreaux (1985) *The Great Merger Movement in American Business, 1895–1904*. New York: Cambridge University Press, pp. 1–2.

4. Daniel Rodgers (1982) "In Search of Progressivism." *Reviews in American History* 10(4): 113–32, p. 126.

5. Dennis Robertson memorably described firms as "islands of conscious power in this ocean of unconscious cooperation, like lumps of butter coagulating in a pail of buttermilk." Dennis Robertson (1923) *The Control of Industry*. London: Nisbet, p. 85.

6. Alfred Chandler (1977) *The Visible Hand*. Cambridge, MA: Harvard University Press, p. 339.

7. *New York Times* (1899) "Opinions of New Yorkers. Albert Shaw and John R. Commons Consider Trusts Inevitable, but Think They Should Be Controlled." *New York Times*, September 14, 1899, p. 2.

8. William F. Willoughby (1898) "The Concentration of Industry in the United States." *Yale Review*, old series, 7:72–92.

9. Rodgers, "In Search of Progressivism," p. 124.

10. See the illustration by G. F. Keller (1882) "The Curse of California." *The Wasp*, August 19, pp. 520–21.

11. Richard T. Ely (1900) *Monopolies and Trusts*. New York: Macmillan, p. 213.

12. Willoughby, "The Concentration of Industry in the United States," p. 88.

13. Walter Lippmann (1914) *Drift and Mastery*. New York: Mitchell Kennerly, p. 124.

14. Herbert Croly (1911 [1909]) *The Promise of American Life*. New York: Macmillan, p. 359.

15. Volney W. Foster (1902) "Organization Is Economy." *Independent* 45(1): 1032–35. Emphasis in original.

16. Allan Nevins (1953) *John D. Rockefeller*, vol. 1. New York: Scribner, p. 402.

17. US Senate (1912) *Report of the Committee on Interstate Commerce, Pursuant to Senate Resolution 98: Hearings on Control of Corporations, Persons and Firms Engaged in Interstate Commerce*, vol. 1, sixty-second Congress, second session. Washington, DC: Government Printing Office, p. 1235.

18. Harold J. Laski (1934) "Mr. Justice Brandeis." *Harper's Magazine*, January, p. 216.

19. Thomas McCraw (1984) *Prophets of Regulation: Charles Francis Adams, Louis D. Brandeis, James M. Landis, Alfred E. Kahn*. Cambridge, MA: Harvard University Press, p. 87.

20. Frederick Winslow Taylor (1911) *The Principles of Scientific Management*. New York: Harper and Brothers.

21. Daniel Nelson cited in Robert Kanigel (1997) *The One Best Way: Frederick Winslow Taylor and the Enigma of Efficiency*. New York: Viking, p. 504.

22. Frank Gilbreth (1912) *Primer on Scientific Management*. New York: D. Van Nostrand, p. 6.

23. Jennifer Karns Alexander (2008) *The Mantra of Efficiency: From Waterwheel to Social Control*. Baltimore: Johns Hopkins University Press, p. 79. To compare 1910 dollars with 2014 dollars, I use a simple arithmetic average of two measures of purchasing power, the Consumer Price Index for urban consumers (CPI-U), which tracks changes in price for a fixed basket of urban consumer goods, and the GDP price deflator, which tracks changes in price for all goods produced, consumer and other. The CPI-U multiplier from 1910 to 2014 is 25.7, and the GDP price deflator multiplier is 18.5. See Samuel H. Williamson (2015) "Seven Ways to Compute the Relative Value of a U.S. Dollar Amount, 1774 to present." *MeasuringWorth*, www.measuringworth.com/uscompare/.

24. Samuel Haber (1964) *Efficiency and Uplift: Scientific Management in the Progressive Era, 1890–1920.* Chicago: University of Chicago Press, p. 148.

25. John R. Commons (1921) *Industrial Government.* New York: Macmillan, p. 272.

26. In Gilbreth, *Primer on Scientific Management,* p. 2.

27. Kanigel, *One Best Way,* pp. 104–5.

28. Ray Stannard Baker (1911) "Frederick W. Taylor—Scientist in Business Management." *American Magazine* 71(5): 564–70.

29. Haber, *Efficiency and Uplift,* p. 94.

30. R. G. Tugwell (1932) "The Principle of Planning and the Institution of Laissez-Faire." *American Economic Review* 2: 75–92, p. 86.

31. Thorstein Veblen (1919) "The Captains of Finance and the Engineers." *Dial,* June 14, pp. 599–606.

32. Taylor, *Principles of Scientific Management,* p. 65.

33. Haber, *Efficiency and Uplift,* p. x.

34. Croly, *The Promise of American Life,* p. 408.

35. Haber, *Efficiency and Uplift,* p. 27.

36. Horace B. Drury (1915) *Scientific Management: A History and Criticism.* PhD dissertation, Columbia University, New York, pp. 138–41.

37. Charles Horton Cooley (1915) "The Progress of Pecuniary Valuation." *Quarterly Journal of Economics* 30 (1): 1–31, p. 20.

38. This motto was inscribed on the cover of the Bureau of Municipal Research's publication, *Municipal Research.*

39. Carl Sandburg (1911) "Making the City Efficient." *LaFollette's Weekly,* September 11, pp. 6–7, 14–15.

40. George B. Hopkins (1912) "The New York Bureau of Municipal Research." *Annals of the American Academy of Political and Social Science* 41: 235–44, p. 235.

41. Chris Nyland (1996) "Taylorism, John R. Commons, and the Hoxie Report." *Journal of Economic Issues* 30(4): 985–1016, p. 992.

42. Axel R Schäfer (2000) *American Progressives and German Social Reform 1875–1920.* Stuttgart: Franz Steiner Verlag.

43. William F. Willoughby (1918) "The Institute for Government Research." *American Political Science Review* 12(1): 49–62.

44. Robert McNutt McElroy (1914) *The Chinese Students' Monthly,* November, pp. 65–67, p. 66.

45. William Franklin Willoughby (1919) "Introduction: The Modern Movement for Efficiency in the Administration of Public Affairs." In Gustavus Adolphus Weber, *Organized Efforts for the Improvement of Methods of Administration in the United States.* Studies in Administration, Institute for Government Research. New York: D. Appleton & Co., p. 4.

46. In Daniel Rodgers (1987) *Contested Truth: Keywords in American Politics since Independence.* New York: Basic Books, p. 182.

47. Woodrow Wilson (1908) *Constitutional Government in the United States.* New York: Columbia University Press, p. 56.

48. Wilson had begun the project as an undergraduate. His graduate advisor, Herbert Baxter Adams, accepted *Congressional Government* in lieu of a PhD dissertation. Had Baxter not been so generous, Woodrow Wilson would not have become the only US president with a PhD.

49. Woodrow Wilson (1956 [1885]) *Congressional Government: A Study in American Politics*. Mineola, NY: Dover Publications, p. 206.

50. Woodrow Wilson, *Constitutional Government in the United States*, p. 70.

51. Woodrow Wilson (1887) "The Study of Public Administration." *Political Science Quarterly* 2(2): 197–222.

52. Robert Hunter (1902) "The Relation Between Social Settlements and Charity Organization." *Journal of Political Economy* 11(1): 75–88, p. 78.

53. Albert Shaw (1909) "The Russell Sage Foundation." *American Review of Reviews* 39(5): 610–11.

54. Ellen Swallow Richards (1910) *Euthenics: The Science of Controllable Environments*. Boston: Whitcomb and Barrows, pp. 152–53.

55. Richards, *Euthenics*, p. vii.

56. Charlotte Perkins Gilman (1913) "The Waste of Private Housekeeping." *Annals of the American Academy of Political and Social Science* 48: 91–95.

57. Irving Fisher (1909) *National Vitality: Its Waste and Conservation*. Washington, DC: Government Printing Office. Pinchot was himself active in the eugenics movement. See Barry Mehler (1988) *A History of the American Eugenics Society, 1921–1940*. PhD dissertation, University of Illinois, Urbana-Champaign, pp. 414–15.

58. Theodore Roosevelt (1913) *Theodore Roosevelt: An Autobiography*. New York: Charles Scribner's Sons, p. 410.

59. *Twentieth Century Magazine* (1912) "A Comparison of the 1912 Platforms." *Twentieth Century Magazine*, October 12, pp. 74–77, p. 74.

60. Fisher, *National Vitality*, pp. 97, 126, 100.

61. Charles Van Hise (1910) *The Conservation of Natural Resources in the United States*. New York: Macmillan, pp. 370, 377, 378. Charles Van Hise (1913) "Address by the President of the University of Wisconsin." *City Club of Philadelphia Bulletin* 6(20): 469–775, p. 471.

62. Roosevelt quoted Patten in his introduction to Charles McCarthy (1912) *The Wisconsin Idea*. With an introduction by Theodore Roosevelt. New York: Macmillan, pp. 28, 29.

63. Residents of Hull-House (1895) *Hull-House Maps and Papers: A Presentation of Nationalities and Wages in a Congested District of Chicago*. New York: Thomas Y. Crowell. See Alice O'Connor (2001) *Poverty Knowledge: Social Science, Social Policy, and the Poor in Twentieth Century U.S. History*. Princeton, NJ: Princeton University Press.

64. See note 23 for method of converting dollar values.

65. Clarke A. Chambers (1971) *Paul U. Kellogg and the Survey: Voices for Social Change and Social Welfare*. Minneapolis: University of Minnesota Press.

66. W.E.B. Du Bois (1899) *The Philadelphia Negro: A Social Study*. Political Economy and Public Law 14. Philadelphia: University of Pennsylvania, p. 3.

67. See O'Connor, *Poverty Knowledge*, pp. 38–39.

68. US Immigration Commission (1911) *Abstracts of Reports of the Immigration Commission*, vol. 1. Washington, DC: Government Printing Office, p. 20.

69. US Immigration Commission, *Abstracts of Reports of the Immigration Commission*, p. 21.

70. See C. Closson (1896) "Social Selection." *Journal of Political Economy* 4(4): 449–66; G. V. de LaPouge and Closson (1897) "The Fundamental Laws of Anthropo-sociology." *Journal of Political Economy* 6(1): 54–92; C. Closson (1899) "The Races of Europe." *Journal of*

Political Economy 8(1): 58–88; C. Closson (1900) "A Critic of Anthropo-sociology." *Journal of Political Economy* 8(3): 397–410; C. Closson (1900) "The Real Opportunity of the So-Called Anglo-Saxon Race." *Journal of Political Economy* 9(1): 76–97. Frank Taussig also published Closson in Harvard's *Quarterly Journal of Economics*. See C. Closson (1896) "Ethnic Stratification and Displacement." *Quarterly Journal of Economics* 11(1): 92–107; C. Closson (1896) "Dissociation by Displacement: A Phase of Social Selection." *Quarterly Journal of Economics* 10(2): 156–86; C. Closson (1900) "Heredity and Environment: A Rejoinder." *Quarterly Journal of Economics* 15(1): 143–46.

71. H.I.B. (1926) "When Ripley Speaks, Wall Street Heeds." *New York Times Magazine*, September 26, pp. 7, 19.

72. Ripley's wrote that his wife's contribution merited co-authorship, but only his name appeared as author. William Z. Ripley (1899) *The Races of Europe: A Sociological Study*. New York: D. Appleton & Co, p. ix.

73. Thorstein Veblen (1912 [1899]) *Theory of the Leisure Class*. New York: Macmillan, pp. 215–20.

74. Veblen returned to race science in "The Mutation Theory and the Blond Race" (1913) and in "The Blond Race and The Aryan Culture" (1913). His last paper was Thorstein Veblen (1934) "An Experiment in Eugenics," published in *Essays in Our Changing Order*. New York: Viking.

75. Ripley, *Races of Europe*, pp. 372–73.

76. William Z. Ripley (1908) "Races in the United States." *Atlantic Monthly*, June, pp. 745–59.

77. *New York Times* (1908) "Future Americans Will Be Swarthy." *New York Times*, November 28, p. 7.

78. LaPouge and Closson, "Fundamental Laws of Anthropo-sociology," p. 54. This story is well told in Terenzio Maccabelli (2008) "Social Anthropology in Economic Literature at the End of the 19th Century: Eugenic and Racial Explanations of Inequality." *American Journal of Economics and Sociology* 67(3): 481–527.

79. Thorstein Veblen (1898) "Why Is Economics Not an Evolutionary Science?" *Quarterly Journal of Economics* 12(4): 373–97.

80. Louis Terman (1916) *The Measurement of Intelligence*. Boston: Houghton Mifflin, pp. 6–7.

81. Terman, *The Measurement of Intelligence*, pp. 6–7.

82. Goddard's biography can be found in Leila Zenderland's fine book on the history of American intelligence testing. Leila Zenderland (1998) *Measuring Minds: Henry Herbert Goddard and the Origins of American Intelligence Testing*. Cambridge: Cambridge University Press.

83. Henry H. Goddard (1917) "Mental Tests and the Immigrant." *Journal of Delinquency* 2(5): 243–77, p. 252, Table II.

84. With test data on well over 1 million recruits, the psychologists believed they now had empirical precision sufficient to scientifically differentiate among the "feeble-minded." Their classification scheme used "moron" to describe adults with the intelligence of a normal child aged eight to twelve. Yerkes and Goddard employed "idiot" to describe adults with a mental age of zero to two, and "imbecile" for adults with a mental age of three to seven. See Zenderland, *Measuring Minds*, pp. 274, 102–3.

85. Robert Yerkes (ed.) (1921) "Psychological Examining in The United States Army." *Memoirs of the National Academy of Sciences*, vol. 15. Washington, DC: Government Printing Office, pp. 789–90.

86. Quoted in Jonathan Spiro (2009) *Defending the Master Race: Conservation, Eugenics and the Legacy of Madison Grant.* Burlington: University of Vermont Press, p. 219.

87. Yerkes, "Psychological Examining in The United States Army," pp. 735–36.

88. Yerkes, "Psychological Examining in The United States Army," pp. 807–8.

89. Richard T. Ely (1918) *The World War and Leadership in a Democracy.* New York: Macmillan, pp. 144–45.

90. Richard T. Ely (1922) "The Price of Progress." *Administration* 3(6): 657–63.

CHAPTER 5: VALUING LABOR: WHAT SHOULD LABOR GET?

1. John Bates Clark (1912) "A Federal Commission on Industrial Relations: Why It Is Needed." *Proceedings of the Academy of Political Science* 2(4): 71–74.

2. See Rosemary Currarino (2011) *The Labor Question in America: Economic Democracy in The Gilded Age.* Urbana: University of Illinois Press.

3. Of course, slaves were valued by the work they could do. The price of a slave measured the value of the work he or she was expected to do. However, slaves could not sell their labor services, because the law denied them the right of self-ownership, and with it the right to alienate their labor for wages.

4. Rogers Smith (1997) *Civic Ideals: Conflicting Visions of Citizenship in U.S. History.* New Haven, CT: Yale University Press.

5. Xenophon's *Oeconomicus* (*oikos* + *nomos),* discusses the *nomos* (laws or rules) of an *oikos* (family, household, or estate).

6. M. I. Finley (1970) "Aristotle and Economic Analysis." *Past and Present* 47(May): 3–25.

7. See Sandra J. Peart and David M. Levy (2008) *The Street Porter and the Philosopher: Conversations on Analytical Egalitarianism.* Ann Arbor: University of Michigan Press.

8. Adam Smith (1904 [1776]). *An Inquiry into the Nature and Causes of the Wealth of Nations by Adam Smith.* Edited with an Introduction, Notes, Marginal Summary and an Enlarged Index by Edwin Cannan. London: Methuen, vol. 2, p. 184.

9. Smith, *Wealth of Nations,* vol. 1, p. 422.

10. Alex Gourevitch (2015) *From Slavery to the Cooperative Commonwealth: Labor and Republican Liberty in the Nineteenth Century.* Cambridge: Cambridge University Press.

11. George M. McNeill (ed.) (1887) *Labor Movement: The Problem of To-day.* Boston: A. M. Bridgeman & Co., p. 411.

12. Daniel T. Rodgers (1978) *The Work Ethic in Industrial America, 1850–1920.* Chicago: University of Chicago Press, pp. 36–37.

13. McNeill, *Labor Movement,* p. 411.

14. Samuel Gompers (1899) "On the Attitude of Organized Labor toward Organized Charity." *American Federationist* 6(4): 79–82, p. 82.

15. E. L. Godkin (1885) "A Just Measure of Wages." *Nation,* January 29, p. 91.

16. Joseph Persky (2000) "The Neoclassical Advent: American Economics at the Dawn of the 20th Century." *Journal of Economic Perspectives* 14(1): 95–108.

17. John Bates Clark (1894) "The Modern Appeal to Legal Forces in Economic Life." *Publications of the American Economic Association* 9(5/6): 9–30, p. 10.

18. Beatrice Webb (1894) "A Living Wage." *Economic Journal* 4(14): 365–68, p. 366.

19. John Bates Clark (1890) "The Law of Wages and Interest." *Annals of the American Academy of Political and Social Science* 1(1): 43–65, p. 44.

20. Clark did not confuse his scientific claim, wages equal marginal product under competitive conditions, with his ethical claim, that workers paid their marginal products get what they deserve. He very plainly understood, as he wrote, "whether labor gets what it produces or not is a question of fact and not of ethics." [John Bates Clark (1899) *Distribution of Wealth: A Theory of Wages, Interest and Profits*. New York: Macmillan, p. 8.]

21. Edward Bemis (1888) "Restriction of Immigration." *Andover Review* 9(March): 251–64, p. 253. Richard T. Ely (1894) "Thoughts on Immigration, No. II." *Congregationalist*, July 5, p. 13. Scott Nearing (1915) "The Inadequacy of American Wages." *Annals of the American Academy of Political and Social Science* 59: 111–24, p. 122.

22. Thomas Stapleford (2009) *The Cost of Living in America: A Political History of Economic Statistics*. Cambridge: Cambridge University Press.

CHAPTER 6: DARWINISM IN ECONOMIC REFORM

1. Edward A. Ross (1891) "Turning toward Nirvana." *Arena*, November, pp. 736–43, p. 739.

2. Simon Nelson Patten (1894) "The Failure of Biologic Sociology." *Annals of the American Academy of Political and Social Science* 4(May): 919–47, p. 924.

3. Irving Fisher (1909) *National Vitality: Its Waste and Conservation*. Washington, DC: Government Printing Office, p. 14.

4. Karl Pearson (1894) "Socialism and Natural Selection." *Fortnightly Review*, July 1, pp. 1–21.

5. Spencer was a man of changing views. Spencer's early radicalism was a far cry from the 1880s and 1890s curmudgeon progressives made their biggest target in the assault on laissez-faire. See Mark Francis (2007) *Herbert Spencer and the Invention of Modern Life*. Ithaca, NY: Cornell University Press.

6. Josiah Strong (1885) *Our Country: Its Possible Future and Its Present Crisis*. New York: Baker & Taylor for the American Home Missionary Society, p. 161.

7. Pyotr Kropotkin (1902) *Mutual Aid: A Factor of Evolution*. London: William Heinemann.

8. David Starr Jordan (1915) *War and the Breed: The Relation of War to the Downfall of Nations*. Boston: Beacon Press.

9. Asa Gray (1861) *Natural Selection Not Inconsistent with Natural Theology: A Free Examination of Darwin's Treatise on the Origin of Species*. London: Trübner & Co.

10. Leslie Jones (1998) "Social Darwinism Revisited." *History Today* 48(8): 7–8.

11. Dorothy Ross (1991) *The Origins of American Social Science*. Cambridge: Cambridge Univerity Press, p. 106.

12. Ernst Mayr (2001) *One Long Argument: Charles Darwin and the Genesis of Modern Evolutionary Thought*. Cambridge, MA: Harvard University Press. I omit discussion of a fifth idea, multiplication of species.

13. Charles R. Darwin (1859) *On the Origin of Species by Means of Natural Selection, or The Preservation of Favoured Races in the Struggle for Life*. London: John Murray, p. 490.

14. This is not to say common descent went uncontested. Polygenism, the theory that different human races were different species created at different times, thus not sharing common descent, had influential partisans in America.

15. The phrase *natura non facit saltum* ("nature doesn't make a leap") appears several times in the *Origin of Species*.

16. "Gene," the modern name given to the basic unit of heredity, was not coined until 1909. Darwin's roughly analogous term was "gemmule," bodies he imagined traveling in the blood, giving rise to the nineteenth-century colloquial expression, "in the blood," as an informal term for "hereditary."

17. Edward J. Larson (2004) *Evolution: The Remarkable History of a Scientific Theory*. New York: Modern Library, p. 86.

18. Darwin died in 1882, too early to witness the debate between the hard hereditarians, who said acquired characters could not be inherited, and the neo-Lamarckians, who said they could.

19. Alfred Russel Wallace (1889) "Letter to the Editor." *Nature* 40(1043): 619. http://people.wku.edu/charles.smith/wallace/S415.htm.

20. The term refers Jean-Baptiste Lamarck (1744–1829), the French naturalist whose name is associated with the idea that traits acquired during an organism's lifetime, can be hereditarily transmitted to progeny.

21. To be clear, Darwinian variation is random, but natural selection is not. Natural selection is blind in the sense that it does not "care" which traits are adaptive. But natural selection does "care" whether a trait is adapted to the individual organism's environment. Darwinian natural selection nonrandomly processes random variation.

22. Darwin, *Origin of Species*, p. 351.

23. Darwin (1871) *Descent of Man, and Selection in Relation to Sex*. London: John Murray, p. 177.

24. Darwin, *Origin of Species*, p. 489.

25. Carl N. Degler (1991) *In Search of Human Nature*. Oxford: Oxford University Press, p. 22.

26. Diane Paul and James Moore (2010) "The Darwinian Context: Evolution and Inheritance." In Alison Bashford and Philippa Levine (eds.), *The Oxford Handbook of the History of Eugenics*. Oxford: Oxford University Press, pp. 27–42, p. 35.

27. Annie L. Cot (2005) "'Breed Out the Unfit and Breed In the Fit': Irving Fisher, Economics, and the Science of Heredity." *American Journal of Economics and Sociology* 64(3): 793–826.

28. Scott Nearing (1911) *Social Sanity: A Preface to the Book of Social Progress*. New York: Moffat, Yard & Co., p. 154.

29. C. R. Henderson (1907) "Are Modern Industry and City Life Unfavorable to the Family?" In *Papers and Proceedings of the First Annual Meeting American Sociological Society, Held at Providence, Rhode Island, December 27–29, 1906*. Chicago: University of Chicago Press, pp. 93–105, p. 103.

30. Lester Frank Ward (1891) "The Transmission of Culture." *Forum* 2(3): 312–19, p. 319.

31. On the persistence of neo-Lamarckian ideas about heredity in American social science, see George Stocking (1962) "Lamarckianism in American Social Science: 1890–1915." *Journal of the History of Ideas* 23(2): 239–56. See also Hamilton Cravens (1978) *The Triumph*

of Evolution: American Scientists and the Heredity-Environment Controversy 1900–1941. Philadelphia: University of Pennsylvania Press.

32. Lester Frank Ward (1898) *Outlines of Sociology.* New York: Macmillan, pp. 257–58.

33. Alfred Russel Wallace (1889) *Darwinism: An Exposition of the Theory of Natural Selection, with Some of Its Applications.* London: Macmillan, p. 40. Darwin, *Origin of Species,* sixth ed., p. 50. Arthur T. Hadley (1907 [1906]) *Standards of Public Morality.* John S. Kennedy Lectures. New York: G. P. Putnam's Sons, pp. 59–60.

34. Eric J. Larson (1997) *Summer for the Gods: The Scopes Trial and America's Continuing Debate over Science and Religion.* Cambridge: Harvard University Press, p. 20.

35. Peter Bowler (1983) *The Eclipse of Darwinism.* Baltimore: Johns Hopkins University Press.

36. Thomas N. Carver (1912) *The Religion Worth Having.* Boston: Houghton Mifflin, p. 88.

37. Lester Frank Ward (1907) "The Establishment of Sociology." *American Journal of Sociology* 12(5): 581–87, p. 585. In *Pure Sociology,* Ward wrote, "war has been the chief and leading condition of human progress" and that "when races stop struggling progress ceases." See Lester Frank Ward (1903) *Pure Sociology.* New York: Macmillan, p. 238. The story of Ward's tortuous attempts to reconcile his race-conflict theory with his opposition to Social Darwinism is well told in Donald C. Bellomy (1984) "'Social Darwinism' revisited." *Perspectives in American History,* new series, 1: 1–129, pp. 54–63.

38. John Bates Clark and John Maurice Clark (1914) *The Control of Trusts.* New York: Macmillan, p. 200.

39. Alfred Russel Wallace (1890) "Human Selection." *Fortnightly Review,* new series, 48: 325–37.

40. James Morone (1998) *The Democratic Wish: Popular Participation and the Limits of American Government.* New Haven, CT: Yale University Press, p. 114. Cushing Strout (1955) "The Twentieth Century Enlightenment." *American Political Science Review* 49(2): 321–39, p. 322.

41. Edward Bellamy (1888) *Looking Backward: 2000–1887.* Boston: Ticknor and Co., pp. 376, 377.

42. On American socialists and evolutionary thought, see Mark Pittenger (1993) *American Socialism and Evolutionary Thought 1879–1920.* Pittsburgh: University of Pittsburg Press.

43. Sidney Webb (1891) "The Difficulties of Individualism." *Economic Journal* 1(2): 360–81, p. 365.

44. William G. Sumner (1914) *The Challenge of the Facts and Other Essays.* Albert Galloway Keller (ed.) New Haven, CT: Yale University Press, p. 90.

45. Edward A. Ross (1903) "Recent Tendencies in Sociology III." *Quarterly Journal of Economics* 17(3): 438–54, p. 447.

46. Charles Horton Cooley (1918) *Social Process.* New York: Charles Scribner's Sons, p. 222.

47. Richard Hofstadter (1944) *Social Darwinism in American Thought, 1860–1915.* Philadelphia: University of Pennsylvania Press.

48. Robert Bannister (1979) *Social Darwinism: Science and Myth in Anglo-American Social Thought.* Philadelphia: Temple University Press. Thomas C. Leonard (2009) "Origins of the Myth of Social Darwinism: The Ambiguous Legacy of Richard Hofstadter's *Social Darwinism in American Thought.*" *Journal of Economic Behavior and Organization* 71: 37–51.

49. Lester Frank Ward (1907) "Social and Biological Struggles." *American Journal of Sociology* 13(3): 289–99, p. 292.

50. Adrian Desmond and James Moore (1991) *Darwin*. London: Michael Joseph.

51. William Coleman (2001) "The Strange 'Laissez Faire' of Alfred Russel Wallace: The Connection between Natural Selection and Political Economy Reconsidered." In J. Laurent and J. Nightingale (eds.), *Darwinism and Evolutionary Economics*. Cheltenham, UK: Edward Elgar, pp. 36–48, p. 39.

52. Donald Winch (2001) "Darwin Fallen among Political Economists." *Proceedings of the American Philosophical Society* 145: 415–37.

53. See Malcolm Rutherford (1998) "Veblen's Evolutionary Programme: A Promise Unfulfilled." *Cambridge Journal of Economics* 22: 463–77. Richard T. Ely (1903) *Studies in the Evolution of Industrial Society*. New York: Macmillan. Simon Nelson Patten (1903) *Heredity and Social Progress*. New York: Macmillan. John R. Commons (1907) *Races and Immigrants*. New York: Macmillan.

54. See Edward A. Ross (1912) *Changing America: Studies in Contemporary Society*. New York: Century; Edward A. Ross (1914) *The Old World in the New: The Significance of Past and Present Immigration to the American People*. New York: Century; and Edward A. Ross (1927) *Standing Room Only*. New York: Century. Charles Richmond Henderson (1893) *Introduction to the Study of the Defective, Dependent and Delinquent*. Boston: D.C. Heath.

55. *Santa Clara County v. Southern Pacific Railroad Company*, 118 US 394 (1886).

56. Henry Carter Adams (1886) "Economics and Jurisprudence." *Science* 8(178): 15–19, p. 17. Herbert Croly (1911 [1909]) *The Promise of American Life*. New York: Macmillan, p. 414. Richard T. Ely (1884) *The Past and the Present of Political Economy*. Johns Hopkins University Studies in Historical and Political Science, second series. Baltimore: N. Murray, p. 49.

57. John R. Commons (1894) *Social Reform and the Church*. New York: Thomas Crowell and Company, pp. 3, 21. Walter Rauschenbusch (1907) *Christianity and the Social Crisis*. New York: Macmillan. Jane Addams (1911) *Twenty Years at Hull House*. New York: Macmillan, p. 124.

58. Richard T. Ely (1889) *Introduction to Political Economy*. New York: Chautauqua Press, p. 92. Edward A. Ross (1901) *Social Control: A Survey of the Foundations of Order*. New York: Macmillan, p. 67.

59. Woodrow Wilson (1918) *The New Freedom*. New York: Doubleday, Page & Co., pp. 47–48.

60. Adams, "Economics and Jurisprudence," p. 16.

61. Lester Frank Ward (1883) *Dynamic Sociology or Applied Social Science as Based upon Statical Sociology and the Less Complex Sciences*. New York: D. Appleton and Co., vol 2, p. 69.

62. Lester Frank Ward (1893) "Psychologic Basis of Social Economics." *Annals of the American Academy of Political and Social Science* 3(January): 72–90.

63. Lester Frank Ward (1883) *Dynamic Sociology or Applied Social Science as Based upon Statical Sociology and the Less Complex Sciences*. New York: D. Appleton and Co., vol. I, p. 662.

64. Croly, *Promise of American Life*, p. 191.

65. Croly, *Promise of American Life*, pp. 399–400.

66. Ely, *Studies in the Evolution of Industrial Society*, pp. 141, 142, 148. Ely (1901) "Social Progress." *Cosmopolitan* 31(1): 61–64, p. 61.

67. Quoted in Yngve Ramstad (1994) "On the Nature of Economic Evolution: John R. Commons and the Metaphor of Artificial Selection." In Lars Magnusson (ed.), *Evolutionary and Neo-Schumpeterian Approaches to Economics*. Boston: Kluwer, p. 65.

68. Commons, *Social Reform and the Church*, pp. 6–7.

69. Ely, *Introduction to Political Economy*, p. 83.

70. Richard A. Gonce (2002) "John R. Commons' 'Five Big Years': 1899–1904." *American Journal of Economics and Sociology* 61(4): 755–77, p. 765.

71. Woodrow Wilson (1889) *The State: Elements of Historical and Practical Politics*. Boston: D.C. Heath & Co., p. 664.

72. Henry C. Adams (1887) "Relation of the State to Industrial Action." *Publications of the American Economic Association* 1(6): 7–85, pp. 41–42.

73. Donald Bellomy (1986) "Two Generations: Modernists and Progressives, 1870–1920." *Perspectives in American History*, new series, 3: 269–306, p. 280.

CHAPTER 7: EUGENICS AND RACE IN ECONOMIC REFORM

1. Carl Kelsey (1909) "Influence of Heredity and Environment upon Race Improvement: An Introductory Paper upon the Significance of the Problem." *Annals of the American Academy of Political and Social Science* 34(1): 3–8.

2. Diane Paul (2001) "History of Eugenics." In Neil Smelser and Paul Baltes (eds.), *International Encyclopedia of Social and Behavioral Sciences*. Amsterdam: Elsevier, pp. 4896–901.

3. Francis Galton (1883) *Inquiries into Human Faculty and Its Development*. London: J. M. Dent and Sons.

4. Francis Galton (1865) "Hereditary Talent and Character." *Macmillan's Magazine* 12:157–66, p. 157.

5. Francis Galton (1904) "Eugenics: Its Definition, Scope and Aims." *American Journal of Sociology* 10(1): 1–25, p. 5.

6. Paul Lombardo (2011) "Introduction: Looking Back at Eugenics." In Paul Lombardo (ed.), *A Century of Eugenics in America: From the Indiana Experiment to the Human Genome Project*. Bloomington: Indiana University Press, p. ix.

7. Paul Lombardo (2003) "Taking Eugenics Seriously: Three Generations of ??? Are Enough." *Florida State University Law Review* 30(2): 191–219, p. 209.

8. Rudloph Vecoli (1960) "Sterilization, a Progressive Measure?" *Magazine of Wisconsin History* 43(3): 190–202.

9. Vecoli, "Sterilization, a Progressive Measure?" p. 196.

10. Edward A. Ross (1936) *Seventy Years of It*. New York: D. Appleton-Century, p. 233.

11. Edward A. Ross (1918) "Introduction." In Paul Popenoe and Roswell Johnson (eds.), *Applied Eugenics*. New York: Macmillan, p. xi.

12. Edward Marshall (1915) "Empty Cradles Worst War Horror: Professor Irving Fisher Says They Will Overshadow Every Other Tragedy of the Conflict." *New York Times*, July 25, section 4, p. 6.

13. William Barber (2005) "Irving Fisher of Yale." *American Journal of Economics and Sociology* 64(1): 43–55, p. 51.

14. Garland Allen (1975) "Genetics, Eugenics and Class Struggle." *Genetics* 74: 33.

15. Daniel Kevles (1995) *In the Name of Eugenics*. Cambridge, MA: Harvard University Press, p. 69.

16. George Hunter (1914) *A Civic Biology: Presented in Problems*. New York: American Book, pp. 261–65, 263.

17. Clarence Darrow, Scopes's defense lawyer, became an outspoken opponent of eugenics. This may be why the defense's evolutionary experts did not appear in court. See Eric Larson (1997) *Summer for the Gods: The Scopes Trial and America's Continuing Debate over Science and Religion*. Cambridge, MA: Harvard University Press.

18. Raymond Pearl (1927) "The Biology of Superiority." *American Mercury*, November, pp. 257–66. See also Clarence Darrow (1926) "The Eugenics Cult." *American Mercury*, June, pp. 129–37.

19. Karl Pearson (1930) *Life, Letters and Labors of Francis Galton*. Cambridge: Cambridge University Press, vol. 3a, p. 220.

20. Francis Galton (1904) "Eugenics: Its Definition, Scope, and Aims." *American Journal of Sociology* 10(1): 1–6. George Bernard Shaw (1904) "Discussion," *American Journal of Sociology* 10(1): 21–23, p. 21.

21. Irving Fisher (1997) *Fisher as Crusader for Social Causes*. The Works of Irving Fisher, vol. 13, William J. Barber (ed.). London: Pickering and Chatto, p. 175. Irving Fisher (1915) "Eugenics—Foremost Plan of Human Redemption." *Proceedings of Second National Conference on Race Betterment*, August 4–8, Battle Creek, MI, pp. 63–66.

22. Irving Fisher (1913) "Eugenics." *Good Health Magazine* 48(11): 561–84, pp. 583–84.

23. Davenport, along with Fisher and John Harvey Kellogg, established the Race Betterment Foundation in 1906. With generous funding from the Harriman railroad fortune, in 1910 Davenport established and directed the Eugenics Record Office of the Cold Spring Harbor Laboratory, which quickly became the center of American eugenic science. For thirty years until his retirement in 1934, Davenport also directed the Station for the Experimental Evolution, funded by the Carnegie Institution and co-located with the Eugenics Record Office. Garland E. Allen (1986) "The Eugenics Record Office at Cold Spring Harbor, 1910–1940: An Essay in Institutional History." *Orisis* 2: 225–64.

24. Christine Rosen (2004) *Preaching Eugenics*. Oxford: Oxford University Press, pp. 93–94.

25. Robert Rydell (1993) "Fitter Families for Future Firesides." In *World of Fairs: The Century of Progress Exhibitions*. Chicago: University of Chicago Press, pp. 38–58. Alexandra Minna Stern (2002) "Making Better Babies: Public Health and Race Betterment in Indiana, 1920–1935." *American Journal of Public Health* 92(5): 742–52.

26. F. Scott Fitzgerald (1925) *The Great Gatsby*. New York: Charles Scriber's Sons, pp. 12–13.

27. Tamsen Wolff (2009) *Mendel's Theatre: Heredity, Eugenics, and Early Twentieth Century American Drama*. New York: Palgrave Macmillan.

28. The examples and quotations of Woolf, Elliot, and Lawrence are drawn from Donald Childs (2001) *Modernism and Eugenics: Woolf, Elliot, Yeats and the Culture of Degeneration*. Cambridge: Cambridge University Press, pp. 10, 23, 79, 90.

29. W. Duncan McKim (1900) *Heredity and Human Progress*. New York: G. P. Putnam's Sons.

30. Two especially useful studies on eugenics in Great Britain and the United States are Daniel Kevles (1995) *In the Name of Eugenics: Genetics and Uses of Human Heredity*. Cambridge, MA: Harvard University Press; and Diane Paul (1995) *Controlling Human Heredity, 1865 to the Present*. Atlantic Highlands, NJ: Humanities Press.

31. See Mark Adams (ed.) (1990) *The Wellborn Science: Eugenics in Germany, France, Brazil, and Russia*. New York: Oxford University Press. For Canada, see A. McLaren (1990)

Our Own Master Race: Eugenics in Canada, 1885–1945. Toronto: McClelland and Stewart. For France, see W. H. Schneider (1990) *Quality and Quantity: The Quest for Biological Regeneration in Twentieth-Century France.* New York: Cambridge University Press. For Latin America, see Nancy Stepan (1991) *The Hour of Eugenics: Race, Gender, and Nation in Latin America.* Ithaca, NY: Cornell University Press. For China, see Frank Dikötter (1992) *The Discourse of Race in Modern China.* London: Hurst; and Frank Dikötter (1998) *Imperfect Conceptions: Medical Knowledge, Birth Defects, and Eugenics in China.* New York: Columbia University Press. For the Scandinavian countries, see Gunnar Broberg and Nils Roll-Hansen (1996) *Eugenics and the Welfare State: Sterilization Policy in Denmark, Sweden, Norway and Finland.* East Lansing: Michigan State University. For Romania, see Maria Bucur (2002) *Eugenics and Modernization in Interwar Romania.* Pittsburgh, PA: University of Pittsburgh Press. For Puerto Rico, see Laura Briggs (2002) *Reproducing Empire.* Berkeley: University of California Press. For Australia, see Diana Wyndham (2003) *Eugenics in Australia: Striving for National Fitness.* London: Galton Institute. For Japan, see Sumiko Otsubo (2005) "Between Two Worlds: Yamanouchi Shigeo and Eugenics in Early Twentieth-Century Japan." *Annals of Science* 62(2): 205–31. For Switzerland, see Véronique Mottier (2008) "Eugenics, Politics and the State: Social Democracy and the Swiss 'Gardening State.'" *Studies in History and Philosophy of Biological and Biomedical Sciences* 39(2): 263–69.

32. Frank Dikötter (1992) *The Discourse of Race in Modern China.* London: Hurst, p. 467.

33. Stepan, *The Hour of Eugenics.*

34. Adams, *The Wellborn Science,* p. 5.

35. *Buck v. Bell* 274 US 200 (1927).

36. Carl N. Degler (1991) *In Search of Human Nature.* Oxford: Oxford University Press, p. 47.

37. Frederick A. Hayek (2011[1960]) "Why I Am Not a Conservative." In Ronald Hamoway (ed.), *The Constitution of Liberty: The Definitive Edition.* Chicago: University of Chicago Press, pp. 519–33.

38. See Jonathan Spiro's excellent account of Grant: Jonathan Spiro (2009) *Defending the Master Race: Conservation, Eugenics and the Legacy of Madison Grant.* Burlington: University of Vermont Press, pp. xii–xxii, 159.

39. It is important to note the unstable ground of these shifting political alliances. Twenty-first-century anti-abortion and anti-birth-control activists condemn Sanger for her post-World War I alliance with American eugenicists. But Sanger herself was anti-abortion, as were many of the eugenics organizations. The American Eugenics Society, in *A Eugenics Catechism,* declared abortion was murder, except to save the life of or prevent serious injury to the mother. See Diane Paul (2014) "What Was Wrong with Eugenics? Conflicting Narratives and Disputed Interpretations." *Science & Education* 23: 259–71, p. 265. See also Kevles, *In the Name of Eugenics,* p. 92.

40. Charles Horton Cooley (1918) *Social Process.* New York: Charles Scribner's Sons, p. 212.

41. Scott Nearing (1911) *Social Sanity: A Preface to the Book of Social Progress.* New York: Moffat, Yard & Co., p. 153.

42. Michael Freeden (1979) "Eugenics and Progressive Thought: A Study in Ideological Affinity." *Historical Journal* 22: 645–71. Mark Pittenger (1993) *American Socialists and Evolutionary Thought, 1870–1920.* Pittsburgh: University of Pittsburg Press.

43. Jennifer Barker-Devine (2009) "'Make Do or Do Without': Women and the Great Depression." In Hamilton Cravens (ed.), *The Great Depression: People and Perspectives*. Santa Barbara, CA: ABC-CLIO, pp. 45–64, p. 52.

44. Wendy Kline (2001) *Building a Better Race.* Berkeley CA: University of California Press, p. 107.

45. David Levy and Sandra Peart have been pioneers in the discovery of the vital connections between eugenic thinking and Anglophone political economy, especially in the nineteenth century. See, for example, Sandra Peart and David Levy (2005) *The "Vanity of the Philosopher": From Equality to Hierarchy in Post Classical Economics.* Ann Arbor: University of Michigan Press; Sandra Peart and David Levy (eds.) (2008) *The Street Porter and the Philosopher: Conversations on Analytical Egalitarianism.* Ann Arbor: University of Michigan Press.

46. John R. Commons (1894) *Social Reform and the Church.* New York: Thomas Crowell and Company, pp. 7, 73, 111–12.

47. Commons, *Social Reform and the Church*, p. 73. Emphasis in original.

48. John R. Commons (1907) *Races and Immigrants.* New York: Macmillan, p. 210.

49. Commons, *Races and Immigrants*, p. 213.

50. The 1890 US Census collected data on mental defectives—the "insane, feeble-minded, deaf and dumb, and blind" (about 500,000)—and on physical defectives—the "sick, deformed, crippled, or otherwise more or less physically disabled" (about 1 million), thus concluding that more than 2 percent of the American population of 62.6 million was defective. John S. Billings (1890) *Report on The Insane, Feeble-Minded, Deaf and Dumb, and Blind in The United States at the Eleventh Census: 1890.* Washington, DC: Government Printing Office.

51. Simon Nelson Patten (1911) "Pragmatism and Social Science." *Journal of Philosophy, Psychology and Scientific Methods* 8(24): 653–60, p. 655.

52. Simon Nelson Patten (1885) *Premises of Political Economy.* Philadelphia: Lippincott, p. 217.

53. Simon Nelson Patten (1911) "The Laws of Environmental Influence." *Popular Science Monthly* 79: 396–402, p. 402.

54. Patten, "Laws of Environmental Influence," p. 402. Simon Nelson Patten (1912) "Types of Men." *Popular Science Monthly* 80: 273–79. Simon Nelson Patten (1915) "Economic Zones and the New Alignment of National Sentiment." *Survey* 3: 612–13, p. 613.

55. Robert Wiebe (1995) *Self Rule: A Cultural History of American Democracy.* Chicago: University of Chicago Press, p. 129.

56. US Immigration Commission (1911) *Dictionary of Races and Peoples. Reports of the Immigration Commission.* Washington, DC: Government Printing Office, vol. 5, p. 3.

57. US Immigration Commission, *Dictionary of Races and Peoples*, vol. 5, p. 3.

58. Edward Saveth (1948) *American Historians and European Immigration 1875–1925.* New York: Columbia University Press.

59. On the views of Progressive Era economists toward African Americans, see William Darity, Jr. (1994) "Many Roads to Extinction: The Early AEA Economists and the Black Disappearance Hypothesis." *History of Economics Review* 21: 47–62. See also Robert Cherry (1976) "Racial Thought and the Early Economics Profession." *Review of Social Economy* 34(2): 147–62; Mark Aldrich (1979) "Progressive Economists and Scientific Racism: Walter Willcox and Black Americans, 1895–1910." *Phylon* 40(1): 1–14.

60. Frederick L. Hoffman (1896) "Race Traits and Tendencies of the American Negro." *Publications of the American Economic Association* 11(1–3): 1–329, pp. 263, 295.

61. Hoffman, "Race Traits and Tendencies of the American Negro," pp. 328–29.

62. Richmond Mayo-Smith (1898 [1890]) *Emigration and Immigration*. New York: Charles Scribner's Sons, pp. 64–65. Richard T. Ely (1898) "Fraternalism vs. Paternalism in Government." *Century* 55(5): 780–84, p. 781.

63. Jackson Lears (2009) *Rebirth of a Nation: The Making of Modern America, 1877–1920*. New York: Harper Collins, p. 98.

64. Commons, *Races and Immigrants*, p. 136.

65. Edward A. Ross (1907) "Comment on D. Collin Wells's 'Social Darwinism.'" *American Journal of Sociology* 12(5): 695–716, p. 695. Edward A. Ross (1901) "Causes of Race Superiority." *Annals of the American Association of Political and Social Science* 18 (July): 67–89, p. 88–89.

66. Charles Horton Cooley (1918) *Social Process*. New York: Charles Scribner's Sons, p. 232.

67. For example, Clark describes early nineteenth-century competition as "economic Darwinism" embodying "the laws of a struggle for existence between competitors." And, "[t]hough the process was savage the outlook which it afforded was not wholly evil. The survival of crude strength was, in the long run, desirable." But Clark referred to competition among firms, not races or persons; he pointedly distinguished the market conditions of the Ricardian era from his own, and he deemed the Ricardian era struggle among firms desirable only because, it "meant to every social class cheapened goods and more comfortable living." John Bates Clark and Franklin Giddings (1888) *The Modern Distributive Process*. Boston: Ginn and Co., p. 2.

68. John Bates Clark (1890) "The Industrial Future of the Negro." In *Mohonk Confernce on the Negro Question*. Boston: George Ellis, pp. 93–96, p. 95.

69. See Robert Prasch (2008) "W.E.B. Du Bois's Contributions to U.S. Economics (1893–1910)." *Du Bois Review* 5(2): 309–24.

70. Kelly Miller (1917) "Eugenics of the Negro Race." *Scientific Monthly* 5(1): 57–59. W.E.B. Du Bois (1903) "The Talented Tenth." In *The Negro Problem*. New York: James Pott & Co., pp. 33–75.

71. W.E.B. Du Bois (1904) "Heredity in the Public Schools." In Eugene F. Provenzo, Jr. (ed.), *Du Bois on Education*. Walnut Creek, CA: AltaMira Press, pp. 111–122, p. 120.

72. My discussion of Du Bois is indebted to Gregory Michael Dorr and Angela Logan (2011) "Quality Not Mere Quantity Counts." In Paul A. Lombardo (ed.), *A Century of Eugenics in America: From the Indiana Experiment to the Human Genome Era*. Bloomington: Indiana University Press, pp. 68–92. The Du Bois quote is from pp. 74–75.

73. Nicole Rafter (1988) *White Trash: The Eugenic Family Studies, 1877–1919*. Boston: Northeastern University Press.

74. Irving Fisher (1921) "Impending Problems of Eugenics." *Scientific Monthly* 13(3): 214–31, p. 225.

75. Edward A. Ross (1903) "Recent Tendencies in Sociology III." *Quarterly Journal of Economics* 17(3): 438–55, p. 447.

76. Edward A. Ross (1901) *Social Control: A Survey of the Foundations of Order*. New York: Macmillan, p. 424. Emphasis in original.

77. Thorstein Veblen (1912 [1899]) *Theory of the Leisure Class*. New York: Macmillan, pp. 215–20.

78. Edward A. Ross (1912) *Changing America: Studies in Contemporary Society*. New York: Century, p. 14.

79. Richard Soloway (1990) *Demography and Degeneration: Eugenics and the Declining Birthrate in Twentieth-Century Britain.* Chapel Hill: University of North Carolina Press, p. 74.

80. David Starr Jordan (1902) *The Blood of the Nation: A Study of the Decay of Nations through the Survival of the Unfit.* Boston: American Unitarian Association, p. 28.

81. Walter Rauschenbusch (1914 [1912]) *Christianizing the Social Order.* New York: Macmillan, p. 90.

82. Rauschenbusch, *Christianizing the Social Order*, p. 178.

83. Rauschenbusch, *Christianizing the Social Order*, pp. 375, 376.

84. Walter Rauschenbusch (1907) *Christianity and the Social Crisis.* New York: Macmillan, p. 275.

85. Rauschenbusch, *Christianizing the Social Order*, p. 41.

86. Willard quoted in Kenneth Rose (1996) *American Women and the Repeal of Prohibition.* New York: New York University Press, p. 26.

87. Eric Kaufman (2004) *The Rise and Fall of Anglo-America.* Cambridge, MA: Harvard University Press, p. 78.

88. Henry Cabot Lodge (1905 [1876]) "The Anglo-Saxon Land-Law." In Henry Adams (ed.), *Essays in Anglo-Saxon Law.* Boston: Little Brown, pp. 55–120.

89. Josiah Strong (1893) *The New Era: Or the Coming Kingdom.* New York: Baker & Taylor, p. 80.

90. Rauschenbusch, *Christianizing the Social Order*, p. 376.

91. By "germs," Adams did not mean infectious microbes, but "germ plasm," the name for the carrier of heredity. Herbert B. Adams (1883) "Saxon Tithingmen in America." *Johns Hopkins University Studies in Historical and Politcal Science* 1(4): 17. Herbert B. Adams (1883) "Norman Constables in America." *Johns Hopkins University Studies in Historical and Political Science* 1(8): 4. Raymond Cunningham (1981) "The German Historical World of Herbert Baxter Adams, 1874–1876." *Journal of American History* 68(2): 269–70.

92. John Burgess (1904) "Germany, Great Britain and the United States." *Political Science Quarterly* 19: 14.

93. Woodrow Wilson (1889) *The State: Elements of Historical and Practical Politics.* Boston: D.C. Heath & Co., p. 2.

94. Wilson, *The State*, pp. 526–27.

95. Wilson (1889) "The Character of Democracy in the United States." *Atlantic Monthly*, November, pp. 577–88, p. 582.

96. G. Stanley Hall (1904) *Adolescence: Its Psychology and Its Relations to Physiology, Anthropology, Sociology, Sex, Crime, Religion and Education.* New York: D. Appleton & Co. Dorothy Ross (1972) *G. Stanley Hall: The Psychologist as Prophet.* Chicago: University of Chicago Press.

CHAPTER 8: EXCLUDING THE UNEMPLOYABLE

1. *Survey* (1914) "Editorial." *Survey*, October 31, pp. 115–16.

2. Simon Nelson Patten (1896) *The Theory of Social Forces.* Philadelphia: American Academy of Political and Social Science, p. 143.

3. Of course, many workers were both—that is, foreign-born women.

4. Charles R. Henderson (1912) "Recent Advances in the Struggle against Unemployment." *American Labor Legislation Review* 2(1): 105–10, p. 107.

5. Sidney Webb and Beatrice Webb (1902) *Industrial Democracy*. London: Longmans, Green, p. 785.

6. See David Moss (1996) *Socializing Security: Progressive-Era Economists and the Origins of American Social Policy*. Cambridge, MA: Harvard University Press, p. 51.

7. Richard T. Ely (1891) "Pauperism in the United States." *North American Review* 152(413): 395–409, p. 407.

8. Ely (1903) *Studies in the Evolution of Industrial Society*. New York: Macmillan, p. 163.

9. Webb and Webb, *Industrial Democracy*, p. 785.

10. Ely, "Pauperism in the United States," p. 395.

11. Walter Lippmann (1915) "The Campaign against Sweating." *New Republic*, March 27, pp. 1–8, p. 7.

12. Franklin Giddings (1919) "The Seven Devils: An Editorial." *Independent* 99(3692): 356–57.

13. *Independent* (1919) "Stopping the Undesirables." *Independent* 99(3692): 352–53.

14. Simon Nelson Patten (1912) "Theories of Progress." *American Economic Review* 2(1): 61–68, p. 64.

15. Samuel Gompers and Herman Guttstadt (1902) *Some Reasons for Chinese Exclusion: Meat vs. Rice: American Manhood against Asiatic Coolieism: Which Shall Survive?* Senate Document 137, fifty-seventh Congress, first session. Washington, DC: Government Printing Office.

16. Woodrow Wilson (1903 [1901]) *A History of the American People*. New York: Harper, vol. V, p. 185.

17. Edward A. Ross (1936) *Seventy Years of It*. New York: D. Appleton-Century, p. 70.

18. John Graham Brooks (1898) "The Trade Union Label." In *Bulletin of the Deptartment of Labor* 15 (March). Washington, DC: Government Printing Office, pp. 197–219, pp. 197–98.

19. John R. Commons and John B. Andrews (1916) *The Principles of Labor Legislation*. New York: Harper and Brothers, p. 74.

20. Quoted in Mary O. Furner (1975) *Advocacy and Objectivity: A Crisis in the Professionalization of American Social Science, 1865–1905*. Lexington: University Press of Kentucky, p. 236.

21. Richard T. Ely (1894) "Thoughts on Immigration, No. II." *Congregationalist*, July 5, p. 13.

22. Richard T. Ely (1894) "Thoughts on Immigration, No. I." *Congregationalist*, June 28, p. 889.

23. Ely, *Studies in the Evolution of Industrial Society*, pp. 165–66.

24. Edward A. Ross (1912) *The Changing Chinese*. New York: Century, p. 47. Ross is cited in Matthew Frye Jacobson (2000) *Barbarian Virtues*. New York: Hill and Wang, p. 77.

25. The promotion of the Irish immigrant to "American" often meant, by the same token, promotion to the white race. Matthew Frye Jacobson (1998) *Whiteness of a Different Color: European Immigrants and the Alchemy of Race*. Cambridge, MA: Harvard University Press.

26. This latter point was emphasized by I. A. Hourwich (1912) *Immigration and Labor: The Economic Aspects of European Immigration to the United States*. New York: G. P. Putnam's Sons.

27. John R. Commons (1907) *Races and Immigrants in America*. New York: Chautauqua Press, pp. 148, 151.

28. Jacob Riis (1997 [1890]) *How the Other Half Lives: Studies among the Tenements of New York*. New York: Penguin Books, p. 92.

29. Frederick Jackson Turner (1901) "Studies of American Immigration." *Chicago Record-Herald*, September 18. Edward N. Saveth (1948) *American Historians and European Immigration 1875–1925*. New York: Columbia University Press, p. 130.

30. Cited in Dave Burns (2008) "The Soul of Socialism: Christianity, Civilization, and Citizenship in the Thought of Eugene Debs." *Labor: Studies in Working-Class History of the Americas* 5(2): 83–116, p. 95. Burns argues that by 1900, Debs had abandoned his nativism and racism, welcoming the Asians and Europeans he once vilified as fellow members of the working class.

31. Nearing assumed that the Lithuanian worker, though equally productive, accepted a wage one-third lower. Scott Nearing (1915) "The Inadequacy of American Wages." *Annals of American Academy of Political and Social Science* 59: 111–24, p. 122.

32. Edward A. Ross (1927) *Standing Room Only*. New York: Century, pp. 318–25. Excerpted in Thomas Elliot (1931) *American Standards and Planes of Living*. Boston: Ginn & Co., p. 608.

33. William Z. Ripley (1904) "Race Factors in Labor Unions." *Atlantic Monthly*, March, pp. 299–308, p. 300.

34. Another machine, the cash register, helped deter the criminals. By forcing cashiers to sound a bell when making change, the machine helped the owner monitor skimming by employees.

35. Hugo Münsterberg (1913) *Psychology and Industrial Efficiency*. Boston: Houghton Mifflin.

CHAPTER 9: EXCLUDING IMMIGRANTS AND THE UNPRODUCTIVE

1. There is a vast literature on immigration and its connection to race, assimilation, and citizenship. I have benefited from the following works. John Higham (1978[1955]) *Strangers in the Land: Patterns of American Nativism*. New York: Atheneum. Rogers Smith (1997) *Civic Ideas: Conflicting Visions of Citizenship in U.S. History*. New Haven, CT: Yale University Press. Matthew Frye Jacobson (1998) *Whiteness of a Different Color: European Immigrants and the Alchemy of Race*. Cambridge, MA: Harvard University Press. Desmond King (2000) *Making Americans: Immigration, Race and the Origins of Diverse Democracy*. Cambridge, MA: Harvard University Press. Gary Gerstle (2001) *American Crucible: Race and Nation in the 20th Century*. Princeton, NJ: Princeton University Press. Daniel Tichenor (2002) *Dividing Lines: The Politics of Immigration Control in America*. Princeton, NJ: Princeton University Press. Mae Ngai (2004) *Impossible Subjects: Illegal Aliens and the Making of Modern America*. Princeton, NJ: Princeton University Press. Robert Zeidel (2004) *Immigrants, Progressives and Exclusion Politics: The Dillingham Commission, 1900–1927*. DeKalb: Northern Illinois University Press.

2. See John Torpey (2000) *The Invention of the Passport: Surveillance, Citizenship and the State*. Cambridge: Cambridge University Press.

3. Claudia Goldin (1994) "The Political Economy of Immigration Restriction in the United States, 1890 to 1921." In Claudia Goldin and Gary Liebcap (eds.), *The Regulation of the Economy*. Chicago: University of Chicago Press, pp. 223–57, p. 239.

4. Jonathan Spiro (2009) *Defending the Master Race: Conservation, Eugenics and the Legacy of Madison Grant*. Burlington: University of Vermont Press, p. 234.

5. Rogers Smith, *Civic Ideals*. Desmond King (1999) *In the Name of Liberalism: Illiberal Social Policy in the USA and Britain*. Oxford: Oxford University Press. Gerstle, *American Crucible*.

6. Leon Czolgosz, a self-professed anarchist, assassinated President William McKinley in September 1901. Though Czolgosz was native-born, the Immigration Act of 1903 barred admission to anarchists. Mormons were the target of the antipolygamy bar.

7. US Immigration Commission (1911) *Abstracts of Reports of the Immigration Commission*. Washington, DC: Government Printing Office, vol. 1, p. 20.

8. US Immigration Commission, *Abstracts of Reports of the Immigration Commission*, p. 21.

9. Richard T. Ely (1910) "The American Economic Association 1885–1909." *American Economic Association Quarterly*, third series, 11(1): 47–111, p. 75.

10. Edward Bemis (1888) "Restriction of Immigration." *Andover Review* 9(March): 251–64, p. 264.

11. Bemis, "Restriction of Immigration," p. 263. The desirable Irish, for Bemis, were the "thrifty and intelligent" Northern Irish, who were ethnically English, and Scottish Protestants, whom he distinguished from the poor and illiterate Catholic Irish.

12. Goldin, "The Political Economy of Immigration Restriction," p. 228. Immigration was procyclical, decreasing during hard times and increasing when demand for labor was strong. So, though "new" immigrants made up a greater share of all immigrants, total immigration declined during the depression-plagued 1890s, rebounding again in the 1900s. Goldin, "The Political Economy of Immigration Restriction," p. 229.

13. Richard T. Ely (1888) *Problems of To-Day*. New York: Thomas Y. Crowell, pp. 76–77.

14. Richmond Mayo-Smith (1898 [1890]) *Emigration and Immigration*. New York: Charles Scribner's Sons.

15. Edwin R. A. Seligman (1919) "Biographical Memoir: Richmond Mayo-Smith, 1854–1901." *Memoirs of the National Academy of Sciences*, vol. 17, second memoir, pp. 71–76.

16. Mayo-Smith, *Emigration and Immigration*, pp. 292, 294.

17. See Dennis Hodgson (1992). "Ideological Currents and the Interpretation of Demographic Trends: The Case of Francis Amasa Walker." *Journal of the History of the Behavioral Sciences* 28(1): 28–44.

18. Francis Amasa Walker (1899) *Discussions in Economics and Statistics*, vol II. Davis R. Dewey (ed.). New York: Henry Holt and Co., p. 447.

19. Walker, *Discussions in Economics and Statistics*, pp. 417–26.

20. Walker, *Discussions in Economics and Statistics*, pp. 422–23. Walker considered and rejected the possibility that fertility decline caused increased immigration; he also rejected the possibility that fertility and immigration were unrelated.

21. E. A. Goldenweiser (1912) "Walker's Theory of Immigration." *American Journal of Sociology* 18(3): 342–51.

22. Walker, *Discussions in Economics and Statistics*, pp. 430, 433.

23. Robert Hunter (1905 [1904]) *Poverty*. New York: Macmillan, pp. 302, 316.

24. Henry Pratt Fairchild (1911) "The Paradox of Immigration." *American Journal of Sociology* 17(2): 254–67, p. 263.

25. Lydia Commander (1907) *The American Idea: Does the National Tendency toward a Small Family Point to Race Suicide or Race Development?* New York: A.S. Barnes & Co., p. 326.

26. Prescott Hall (1904) "Selection of Immigration." *Annals of the American Academy of Political and Social Science* 19: 169–84, p. 182.

27. Race-suicide theorists abroad differed only on the matter of which inferiors constituted the eugenic threat. Sidney Webb described race suicide in England this way: "Twenty-five percent of our parents, as Professor Karl Pearson keeps warning us, is producing 50 percent of the next generation. This can hardly result in anything but national deterioration; or, as an alternative, in this country gradually falling to the Irish and the Jews." Sidney Webb (1907) *The Decline in the Birth-Rate*. Fabian Society Tract 131. London: Fabian Society, p. 17.

28. Charles Darwin (1898) *The Descent of Man and Selection in Relation to Sex*, second ed. New York: D. Appleton & Co., pp. 144–45.

29. At the same moment Turner said the race-invigorating American frontier had disappeared, Alfred Marshall, the leading English economist of the day, also used natural selection to explain the character of a nation's people. Writing in *Principles of Economics* (1890), the canonical Anglophone economics textbook for a generation, Marshall said "England's geographical position caused her to be peopled by the strongest members of the strongest races of northern Europe; a process of natural selection brought to her shores those members of each successive migratory wave who were most daring and self-reliant."

30. John R. Commons (1907) *Races and Immigrants in America*. New York: Chautauqua Press, pp. 127–28.

31. Edward T. Devine (1911) "Selection of Immigrants." *Survey*, February 4, pp. 715–16.

32. Edward A. Ross (1904) "The Value Rank of the American People." *Independent* 57 (November): 1061–64, p. 1063.

33. John R. Commons (1901) "Immigration and Its Economic Effects." In *Reports of the U.S. Industrial Commission*. Washington, DC: Government Printing Office, vol. 15, pp. 293–743.

34. Commons, "Immigration and Its Economic Effects," p. 327.

35. Commons, *Races and Immigrants*, p. 153.

36. John R. Commons (1907) "Review of *Immigration and Its Effect upon the United States* by Prescott F. Hall." *Charities and the Commons* 17: 504.

37. Barbara M. Solomon (1956) *Ancestors and Immigrants*. Chicago: University of Chicago Press, p. 132.

38. Ross, *The Old World in the New*, p. 286.

39. Quotes are from Edward A. Ross (1901) "Causes of Race Superiority." *Annals of the American Association of Political and Social Science* 18(July): 67–89, pp. 83, 84, 88–89; Ross, *Old World in the New*, pp. 136, 219, 254, 285, 286, 291; Edward A. Ross (1921) "The Menace of Migrating People." *Century* 102 (May): 131–35, p. 134.

40. In Mark H. Haller (1984) *Eugenics: Hereditarian Attitudes in American Thought*. New Brunswick, NJ: Rutgers University Press, p. 144. Irving Fisher (1921) "Impending Problems of Eugenics." *Scientific Monthly* 13(3): 214–31, pp. 226–27.

41. Charles R. Henderson (1909) "Are Modern Industry and City Life Unfavorable to the Family?" *American Economic Association Quarterly*, third series, 10(1): 217–32, p. 232.

42. Charles R. Henderson (1900) "Science in Philanthropy." *Atlantic Monthly*, February, pp. 249–54, p. 253.

43. Charles A. Ellwood (1910) *Sociology and Modern Social Problems.* New York: American Book, pp. 188–90.

44. Theodore Roosevelt (1907) "A Letter from President Roosevelt on Race Suicide." *American Monthly Review of Reviews* 35(5): 550–51.

45. The Commission recommended seven possible methods of exclusion: the literacy test, race-based quotas, barring unskilled laborers "unaccompanied by wives or families," limits on total immigration, a wealth test, an increase in the head tax, and an increase in head tax on men without families. US Immigration Commission, *Abstracts of Reports of the Immigration Commission*, pp. 47–48.

46. Jeremiah Jenks and W. Jett Lauck (1913) *The Immigration Problem*, third ed. New York: Funk and Wagnalls, p. xx.

47. US Immigration Commission, *Abstracts of Reports of the Immigration Commission*, p. 14. Oscar Handlin (1957) *Race and Nationality in American Life*. Garden City, NY: Doubleday Anchor, pp. 80–81.

48. US Immigration Commission, *Abstracts of Reports of the Immigration Commission*, p. 14.

49. Tichenor, *Dividing Lines*, pp. 130–31.

50. Tichenor, *Dividing Lines*, p. 129, citing Solomon, *Ancestors and Immigrants*, pp. 128, 152.

51. Jeanne D. Petit (2010) *The Men and Women We Want: Gender, Race, and the Progressive Era Literacy Test Debate.* Rochester, NY: University of Rochester Press, p. 160, n. 24.

52. Zeidel, *The Dillingham Commission*, p. 105.

53. US Immigration Commission, *Abstracts of Reports of the Immigration Commission*, p. 18.

54. Woodrow Wilson (1889) *The State: Elements of Historical and Practical Politics*. Boston: D.C. Heath & Co., p. 475.

55. See, for example, Commons, *Races and Immigrants*, pp. 16–18.

56. Folkmar made clear that the dictionary was not intended as an original contribution to ethnology but as a distillation of the race science literature, a kind of field guide for the "student of immigration." US Immigration Commission, *Dictionary of Races and Peoples*, p. 3.

57. Zeidel, *The Dillingham Commission*, p. 106.

58. US Immigration Commission, *Dictionary of Races*, pp. 23, 82, 83, 126, 66, 47, 100.

59. Jenks and Lauck, *The Immigration Problem*, p. 369.

60. Jeremiah Jenks (1913) "The Character and Influence of Recent Immigration." In *Questions of Public Policy*. New Haven, CT: Yale University Press, pp. 1–40, p. 6.

61. Jenks and Lauck, *The Immigration Problem*, p. 370.

62. Jenks and Lauck, *The Immigration Problem*, p. 370.

63. Ross, *The Old World in the New*, p. 165.

64. John Dewey (1922) "Racial Prejudice and Friction." *Chinese Social and Political Science Review* 6: 1–17, pp. 11, 13–14. See Gary Gerstle (1994) "The Protean Character of American Liberalism." *American Historical Review* 99(4): 1043–73, pp. 1058–59.

65. Cited in Axel R. Schäfer (2000) *American Progressives and German Social Reform, 1875–1920*. Stuttgart: Franz Steiner Verlag, p. 178, n. 138.

66. See Florence Kelley (2009) *The Selected Letters of Florence Kelley, 1869–1931*. Edited by Katherine Kish Sklar and Beverly Wilson Palmer. Urbana: University of Illinois Press, p. 169.

67. Selig Perlman (1918) "The Anti-Chinese Agitation in California." In John R. Commons, Helen L. Sumner, David Joseph Saposs, John B. Andrews, Selig Perlman, and Henry Elmer Hoagland, *History of Labor in the United States*. New York: Macmillan, vol. 2, pp. 252–68, p. 252.

68. Jeremiah Jenks (1913) "The Character and Influence of Recent Immigration." In *Questions of Public Policy*. New Haven CT: Yale University Press, pp. 1–40, p. 2.

69. Franz Boas (1912) "Changes in the Bodily Form of Descendants of Immigrants." *American Anthropologist*, new series, 14(3): 530–62, p. 530.

70. Boas, "Changes in the Bodily Form of Descendants of Immigrants," p. 550. Boas noted that his report used the term "types," but the Dillingham Commission editors replaced it with "races."

71. Jonathan Spiro (2009) *Defending the Master Race: Conservation, Eugenics and the Legacy of Madison Grant*. Burlington: University of Vermont Press, p. 298.

72. Horace M. Kallen (1915) "Democracy versus the Melting-Pot: A Study of American Nationality." *Nation*, February 18 and February 25, pp. 190–94, 217–20.

73. US House of Representatives (1901) *US Industrial Commission Reports*, vol. XV, Document 184, fifty-seventh Congress, first session. Washington, DC: Government Printing Office, pp. 66–67.

74. Solomon, *Ancestors and Immigrants*, p. 180, citing Edward Atkinson (1892) "Incalculable Room for Immigrants." *Forum* 13: 360–70.

75. On the diversity and membership of the Progressive Era coalitions, see Tichenor, *Dividing Lines*, p. 121.

76. Emily Greene Balch (1910) *Our Slavic Fellow Citizens*. New York: Charities Publication Committee, pp. 286–87.

77. She also publicly opposed American involvement in the First World War, a courageous stance that cost her the Political Economy chair at Wellesley but later was cited when she was awarded the Nobel Peace Prize in 1946.

78. Emily Greene Balch (1910) "The Education and Efficiency of Women." *Proceedings of the Academy of Political Science in the City of New York* 1(1): 61–71, p. 63.

79. Frank A. Fetter, William B. Bailey, Henry C. Potter, Emily Greene Balch, I. M. Rubinow, C.W.A. Veditz, and Walter E. Willcox (1907) "Western Civilization and Birth-Rate—Discussion." *Publications of the American Economic Association*, third series, 8(1): 90–112, p. 102.

80. Goldin, "The Political Economy of Immigration Restriction," p. 256. Grover Cleveland vetoed the 1897 literacy-test bill. William Howard Taft vetoed the 1913 bill, and Woodrow Wilson vetoed the 1915 bill. Congress overrode a fourth presidential veto, the second by Wilson, to pass the literacy test in 1917.

81. Woodrow Wilson (1903) *History of the American People*, vol V. New York: Harper and Brothers, pp. 212–13.

82. Higham, *Strangers in the Land*, p. 190.

83. Edward A. Ross (1914) *The Old World in the New: The Significance of Past and Present Immigration to The American People*. New York: Century, pp. 144–45, 154.

84. Edward T. Devine (1911) "Selection of Immigration." *Survey*, February 4, pp. 715–16.

85. Daniel M. Fox (1968) "Introduction." In Simon Nelson Patten, *The New Basis of Civilization*. Cambridge, MA: Belknap Press, pp. xxxiii–xxxiv.

86. Devine, "Selection of Immigration," pp. 715–16.

87. Ely, *Problems of To-Day*, p. 74.

88. Ross was quoted in the *San Francisco Chronicle*, May 8, 1900, cited in Roger Daniels (1988) *Asian America: Chinese and Japanese in the United States Since 1850*. Seattle: University of Washington Press, p. 112.

89. US Bureau of the Census (1975) *Historical Statistics of the United States, Colonial Times to 1970*, vol. I. Series D 778. Washington, DC: Government Printing Office, p. 168.

90. Paul Kellogg (1913) "Immigration and the Minimum Wage." *Annals of the American Academy of Political and Social Science* 48(July): 66–77, p. 75.

91. Frank Taussig (1916) "Minimum Wages for Women." *Quarterly Journal of Economics* 30(3): 411–42, p. 426.

92. John A. Ryan (1917) "Minimum Wage Legislation." In Mary Katherine Reely (ed.), *Selected Articles on Minimum Wage*. New York: H. W. Wilson, p. 41. Sidney Webb (1912) "The Economic Theory of a Legal Minimum Wage." *Journal of Political Economy* 20(10): 973–98.

93. Thomas C. Leonard (2000) "The Very Idea of Applied Economics: The Modern Minimum-Wage Controversy and Its Antecedents." In Roger Backhouse and Jeff Biddle (eds.), *Toward a History of Applied Economics: History of Political Economy* 32(suppl.): 117–44.

94. Henry Sidgwick (1886) "Economic Socialism." *Contemporary Review* 50 (November): 620–31. Alfred Marshall (1897) "The Old Generation of Economists and the New." *Quarterly Journal of Economics* 11(2): 115–35. Arthur Cecil Pigou (1913) "The Principle of the Minimum Wage." *Nineteenth Century* 73: 644–58. Philip H. Wicksteed (1913) "The Distinction between Earnings and Income, and between a Minimum Wage and a Decent Maintenance: A Challenge." In *The Industrial Unrest and the Living Wage*. London: Collegium, pp. 76–86.

95. The Fabian Society wrote (Tract 128, p. 12), "it must not be assumed that all the workers who are now receiving less than the legal minimum would be thrown out of employment. But some of them would."

96. Sidney Webb (1912) "Economic Theory of a Legal Minimum Wage." *Journal of Political Economy* 20(10): 973–98, p. 993.

97. C. R. Henderson (1898) "Review of *Industrial Democracy* by Sidney Webb, Beatrice Webb." *American Journal of Sociology* 3(6): 850–55.

98. Charles Richmond Henderson (1909) "Are Modern Industry and City Life Unfavorable to the Family?" *American Economic Association Quarterly*, third series, 10(1): 217–32, p. 224.

99. Webb and Webb, *Industrial Democracy*, p. 785.

100. Henderson, "Are Modern Industry and City Life Unfavorable to the Family?" pp. 228–29.

101. Henry R. Seager (1913) "The Minimum Wage as Part of a Program for Social Reform." *Annals of the American Academy of Political and Social Science* 48 (July): 3–12. Henry R. Seager (1913) "The Theory of the Minimum Wage." *American Labor Legislation Review* 3: 81–91, pp. 81, 82–83.

102. Seager, "The Minimum Wage as Part of a Program for Social Reform," p. 9.

103. Seager (1917) *Principles of Economics*, second ed. New York: Henry Holt, p. 586.

104. Seager, "The Minimum Wage as Part of a Program for Social Reform," pp. 9–10.

105. Seager, *Principles of Economics*, p. 310.

106. A. B. Wolfe, Robert L. Hale, and John A. Ryan (1917) "Some Phases of the Minimum Wage: Discussion." *American Economic Review* 7: 275–81, pp. 275, 278.

107. Webb and Webb, *Industrial Democracy*, p. 787.

108. Sidney Ball (1896) "The Moral Aspects of Socialism." *International Journal of Ethics* 6(3): 290–322: pp. 290, 295.

109. A. N. Holcombe (1912) "The Legal Minimum Wage in the United States." *American Economic Review* 2(1): 21–37.

110. Edward Cummings (1899) "A Collectivist Philosophy of Trade Unionism." *Quarterly Journal of Economics* 13(2): 151–86, p. 178.

111. Thomas Reed Powell (1917) "The Oregon Minimum-Wage Cases." *Political Science Quarterly* 32(2): 296–311, pp. 296, 310.

112. Royal Meeker (1910) "Book Review: *Cours d'Economie Politique*." *Political Science Quarterly* 25(3): 543–45, p. 544.

113. Paul Popenoe and Roswell Johnson (1920) *Applied Eugenics*. New York: Macmillan, pp. 374–75.

114. Frank A. Fetter (1907) "Western Civilization and the Birthrate." *Publications of the American Economic Association*, third series, 8: 92–93.

115. Frank A Fetter (1915) *Principles of Economics*. New York: Century, pp. 421–22.

116. Ann T. Keene (2000) "Taussig, Frank William." In American National Biography Online. http://www.anb.org/articles/14/14–00620.html.

117. Frank Taussig (1916) "Minimum Wages for Women." *Quarterly Journal of Economics* 30(3): 411–42, p. 422.

118. Frank Taussig (1911) *Principles of Economics*, vol 11. New York: Macmillan, p. 300.

119. Scott Nearing (1911) "'Race Suicide' vs. Overpopulation." *Popular Science Monthly*, January, pp. 81–83. Scott Nearing (1914) "The Geographical Distribution of American Genius." *Popular Science Monthly*, August, pp. 189–99. Scott Nearing (1916) "The Younger Generation of American Genius." *Scientific Monthly* 2(1): 48–61. Nellie Nearing and Scott Nearing (1912) "What Is Meant by Eugenics." *Ladies' Home Journal*, September, p. 14.

120. Scott Nearing (1911) *Social Sanity: A Preface to the Book of Social Progress*. New York: Moffat, Yard & Co., p. 153.

121. Scott Nearing (1912) *The Super Race: An American Problem*. New York: B. W. Huebsch, pp. 31–32.

122. Nearing, *The Super Race*, pp. 39, 40.

123. Another example concerns Irving Fisher. In his *Elementary Principles*, he declared that "if the vitality or vital capital is impaired by a breeding of the worst and a cessation of the breeding of the best, no greater calamity could be imagined." Fortunately, said Fisher, eugenics offered a means, "by isolation in public institutions and in some cases by surgical operation," to prevent the calamity of "inheritable taint." Irving Fisher (1912) *Elementary Principles of Economics*. New York: Macmillan, p. 476.

CHAPTER 10: EXCLUDING WOMEN

1. Elizabeth Brandeis (1966 [1935]) "Labor Legislation." In *History of Labor in the United States, 1896–1932*. New York: Augustus M. Kelley, vol. III, pp. 399–700, pp. 459, 501. J. Stanley Lemons (1988) "Mothers' Pensions Acts." In John D. Buenker and Edward R.

Kantowicz (eds.). *Historical Dictionary of the Progressive Era, 1890–1920.* Westport, CT: Greenwood Press, pp. 291–92.

2. Male public-works and railroad workers were an exception, created by the Eight Hours on Public Works Act of March 3, 1913, and the Adamson Act, September 3 and 5, 1916. Generally, courts permitted labor regulation to infringe on men's constitutional right to free contract only in cases of hazard—in dangerous occupations, such as coal mining, or where tired employees presented a danger to others, as with railroad or streetcar workers. *Holden v. Hardy* 169 US 366, 398 (1898).

3. Robyn Muncy (1999) "The Ambiguous Legacies of Women's Progressivism." *OAH Magazine of History* 13(3): 15–19, pp. 17–18.

4. Josephine Goldmark (1912) *Fatigue and Efficiency: A Study in Industry.* New York: Charities Publication Committee.

5. Richard T. Ely (1894) *Socialism and Social Reform.* New York: Thomas Crowell and Company, p. 322.

6. Goldmark, *Fatigue and Efficiency*, p. 39.

7. John Louis Recchiuti (2007) *Civic Engagement: Social Science and Progressive-Era Reform in New York City.* Philadelphia: University of Pennsylvania Press, pp. 135, 136.

8. Henry R. Seager (1913) "The Minimum Wage as Part of a Program for Social Reform." *Annals of the American Academy of Political and Social Science* 48 (July): 3–12, p. 11.

9. John Bates Clark (1913) "The Minimum Wage." *Atlantic Monthly* 112 (September): 289–297, p. 294.

10. Sophonisba P. Breckinridge (1906) "Legislative Control of Women's Work." *Journal of Political Economy* 14(2): 107–9, p. 108.

11. Florence Kelley (1911) "Minimum Wage Boards." *American Journal of Sociology* 17(3): 303–14, p. 304.

12. Florence Kelley (1905) *Some Ethical Gains through Legislation.* New York: Macmillan, p. 142.

13. Kelley, "Minimum Wage Boards," p. 304.

14. Jane Humphries (1976) "Women: Scapegoats and Safety Values in the Great Depression." *Review of Radical Political Economics* 8(1): 98–121, p. 105.

15. Florence Kelley (1914) *Industry in Relation to the Family, Health, Education, Morality.* New York: Longmans Green, p. 16.

16. Florence Kelley (1912) "Minimum-Wage Laws." *Journal of Political Economy* 20(10): 999–1010: p. 1003.

17. Cited in Robyn Muncy (1991) *Creating a Female Dominion in American Reform, 1890–1935.* Oxford: Oxford University Press, p. 162.

18. Theresa McMahon (1925) *Social and Economic Standards of Living.* Boston: D.C. Heath.

19. Seager, "Minimum Wage as Part of a Program for Social Reform," p. 4.

20. Richard T. Ely (1893) "Introduction." In Helen Campbell (1893) *Women Wage-Earners: Their Past, Their Present, and Their Future.* Boston: Roberts Brothers, p. vi.

21. Cited in Daniel Tichenor (2002) *Dividing Lines: The Politics of Immigration Control in America.* Princeton, NJ: Princeton University Press, p. 122.

22. Helen Sumner (1910) *Report on Condition of Woman and Child Wage-Earners in the United States. Volume IX: History of Women in Industry in the United States.* Sixty-first Congress, second session. Washington, DC: Government Printing Office, p 11.

23. Daniel Hammond (2011) "Strange Bedfellows: Fr. John A. Ryan and the Minimum Wage Movement." *Journal of the History of Economic Thought* 33(4): 449–66.

24. John A. Ryan (1906) *A Living Wage: Its Ethical and Economic Aspects*. New York: Macmillan, p. 119.

25. John A. Ryan (1920) *Social Reconstruction*. New York: Macmillan, p. 65.

26. Ryan, *A Living Wage*, pp. 107, 120.

27. Ryan, *A Living Wage*, p. 133.

28. William Hard and Rheta Childe Dorr (1908) "The Woman's Invasion, part II." *Everyone's Magazine*, November, pp. 798–810, p. 798. The charge that the "most objectionable feature" of woman in industry was her "irresponsible cheapness" was added by the magazine's editor, in an brief introduction to Hard and Dorr's six-part series.

29. Hard and Dorr, "The Woman's Invasion, part II," p. 798.

30. Walter Lippmann (1915) "The Campaign against Sweating." *New Republic*, March 27, pp. 1–8.

31. Vivien Hart (1994) *Bound by Their Constitution*. Princeton, NJ: Princeton University Press, p. 93.

32. Edward A. Ross (1912) *Changing America: Studies in Contemporary Society*. New York: Century, p. 77.

33. See Ellen Fitzpatrick (1990) *Endless Crusade: Women Social Scientists and Progressive Reform*. New York: Oxford University Press.

34. Sophonisba Breckenridge (1923) "The Home Responsibilities of Women Workers and the 'Equal Wage.'" *Journal of Political Economy* 31(4): 521–43.

35. Breckenridge, "The Home Responsibilities of Women Workers."

36. Linda Gordon (1992) "Social Insurance and Public Assistance: The Influence of Gender in Welfare Thought in the United States, 1890–1935." *American Historical Review* 97(1): 19–54, pp. 19, 47.

37. *New York Times* (1912) "What the Study of the Relation of Fatigue and Efficiency Shows." *New York Times*, July 7.

38. Irving Fisher (1909) *National Vitality: Its Waste and Conservation*. Washington, DC: Government Printing Office, pp. 4–5.

39. Felix Frankfurter, Mary W. Dewson, and John R. Commons (1924) *State Minimum Wage Laws in Practice*. New York: National Consumers' League, p. 113.

40. *Muller v. Oregon*, 208 US 412, 422 (1908) (USSC+).

41. *New York Times* (1908) "Penalty Woman Pays to Industrial Progress." *New York Times*, May 3, p. 50.

42. Charlotte Perkins Gilman (1900) *Concerning Children*. Boston: Small, Maynard & Co., p. 244.

43. Charlotte Perkins Gilman (1898) *Women and Economics: A Study of the Economic Relation between Men and Women as a Factor in Social Evolution*. Boston: Small, Maynard & Co., pp. viii, 46.

44. Mary Ziegler (2008) "Eugenic Feminism: Mental Hygiene, the Women's Movement, and the Campaign for Eugenic Legal Reform, 1900–1935." *Harvard Journal of Law and Gender* 31: 211–35.

45. Theodore Roosevelt (1911) "Race Decadence." *Outlook*, April 8, pp. 763–68, p. 767.

46. Theodore Roosevelt (1914) "Twisted Eugenics." *Outlook*, January 3, pp. 30–34, p. 32.

47. Albert Benedict Wolfe (1906) *The Lodging House Problem in Boston*. Boston: Houghton Mifflin, pp. 164–65.

48. Simon Patten (1910) "The Crisis in American Home Life." *Independent*, February 17, pp. 342–46, pp. 344–45, 346.

49. Linda Gordon (2007) *The Moral Property of Women: A History of Birth Control Politics in America*. Urbana: University of Illinois Press, p. 88.

50. G. Stanley Hall (1908) "The Kind of Women Colleges Produce." *Appleton's Magazine*, September, pp. 313–19. J. C. Phillips (1916) "A Study of the Birthrate in Harvard and Yale Graduates." *Harvard Graduates Magazine*, September, pp. 25–34. Kelly Miller (1917) "Eugenics of the Negro Race." *Scientific Monthly* 5(1): 57–59. Mary Van Kleeck (1918) "Census of College Women." *Journal of the Association of Collegiate Alumnae* 11(1): 557–91.

51. Nellie S. Nearing (1914–1915) "Education and Fecundity." *Quarterly Publications of the American Statistical Association* 14: 156–74. Irving Fisher (1921) "Impending Problems of Eugenics." *Scientific Monthly* 13(3): 214–231, p. 225.

52. Fisher, "Impending Problems of Eugenics," p. 225.

53. Gordon, *The Moral Property of Women*, p. 89.

54. Scott Nearing and Nellie Nearing (1912) *Woman and Social Progress*. New York: Macmillan, p. 17.

55. See, for example, Hart, *Bound by Our Constitution*.

56. See Muncy, *Creating a Female Dominion*.

57. The previous two paragraphs are indebted to an excellent survey: Patrick Wilkinson (1999) "The Selfless and the Helpless: Maternalist Origins of the U.S. Welfare State." *Feminist Studies* 25(3): 571–97.

58. *Adkins v. Children's Hospital*, 261 US 525, 558–9 (1923) (USSC+).

59. *Adkins v. Children's Hospital*, 261 US 525, 562, 567, 570 (1923) (USSC+).

60. Henry R. Seager (1923) "The Minimum Wage: What Next?" *Survey*, May 15, pp. 215–16, p. 215.

61. On the aftermath of Adkins among labor reformers, see Hart's excellent *Bound by Our Constitution*, pp. 130–143, from which my account draws.

62. *New York Times* (1920) "Women's Work Limited by Law." *New York Times*, January 18, p. 94.

EPILOGUE

1. Dorothy Ross (1984) "American Social Science and the Idea of Progress." In Thomas L. Haskell (ed.), *The Authority of Experts*. Bloomington: Indiana University Press, pp. 157–75, pp. 157–58.

2. Irving Fisher (1919) "Economists in Public Service." *American Economic Review* 9(1): 5–21, p. 7.

3. Herbert Croly (1924) "Introduction." In E. C. Lindeman, *Social Discovery*. New York: Republic, pp. ix–xx, p. xii.

4. Croly, "Introduction," pp. xii, xiv.

5. Croly, "Introduction," p. xiii.

6. Sidney Webb (1910–1911) "Eugenics and the Poor Law: The Minority Report." *Eugenics Review* 2(3): 233–41, p. 237.

7. Francis Galton (1904) "Eugenics: Its Definition, Scope and Aims." *American Journal of Sociology* 10(1): 1–25, p. 5.

8. John Searle (1998) "The Early Years." In R. A. Peel (ed.), *Essays in the History of Eugenics*. London: Galton Institute.

9. Edward A. Ross (1918) "Introduction." In Paul Popenoe and Roswell Johnson (eds.), *Applied Eugenics*. New York: Macmillan, p. xi.

10. Jane Addams (1912) *A New Conscience and an Ancient Evil*. New York: Macmillan, pp. 130–33.